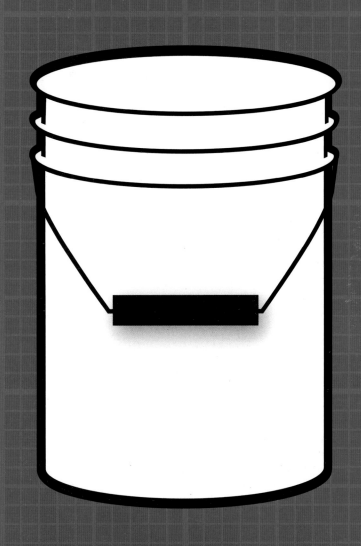

Quarto is the authority on a wide range of topics.

Quarto educates, entertains and enriches the lives of our readers—enthusiasts and lovers of hands-on living.

www.quartoknows.com

First published in 2015 by Voyageur Press an imprint of Quarto Publishing Group USA Inc., 400 First Avenue North, Suite 400, Minneapolis, MN 55401 USA.
Telephone: (612) 344-8100 Fax: (612) 344-8692

quartoknows.com
Visit our blogs at quartoknows.com

Voyageur Press titles are also available at discounts in bulk quantity for industrial or sales-promotional use. For details contact the Special Sales Manager at Quarto Publishing Group USA Inc., 400 First Avenue North, Suite 400, Minneapolis, MN 55401 USA.

10 9 8 7 6 5

ISBN: 978-0-7603-4789-8

Library of Congress Cataloging-in-Publication Data

Peterson, Chris, 1961-
  The 5-gallon bucket book : useful DIY projects, hacks and upcycles / Chris Peterson.
     pages cm
  Includes index.
  ISBN 978-0-7603-4789-8 (sc)
  1. Plastics craft. 2. Recreation--Equipment and supplies--Design and construction. 3. Implements, utensils, etc.--Design and construction. 4. Garden ornaments and furniture--Design and construction. 5. Pails. 6. Salvage (Waste, etc.) I. Title. II. Title: Five-gallon bucket book.
  TT297.P48 2015
  745.57'2--dc23
                              2015022632

Acquiring Editor: Thom O'Hearn
Project Manager: Madeleine Vasaly
Art Director: Cindy Samargia Laun
Book Design: Carol Holtz
Layout: Simon Larkin

On the front cover: Upside-Down Tomato Planter, page 124
On the back cover: From left to right, Chicken Feeder, page 72; Cyclone Dust Collector, page 34; Toddler's Swing, page 98; and Rod-and-Tackle Bucket, page 57

Printed in China

5-GALLON BUCKET BOOK

DIY Projects, Hacks, and Upcycles

Chris Peterson

Voyageur Press

CONTENTS

HANDY HOME ADDITIONS

BARNYARD HELPERS

FAMILY FUN

YARD AND GARDEN INNOVATIONS

Introduction

THE FIVE-GALLON BUCKET IS AMONG THE BEST OF MAN'S INVENTIONS, RIGHT UP THERE WITH THE WHEEL AND SLICED BREAD. NOT ONLY ARE THESE HANDLED CONTAINERS THE PERFECT SIZE FOR CARRYING A WEALTH OF DIFFERENT MATERIALS FROM ONE PLACE TO ANOTHER, THEY CAN BE UPCYCLED INTO A MIND-BOGGLING ARRAY OF NEW CONVENIENCES AND HANDY INVENTIONS.

That's because these buckets are both remarkably strong and surprisingly adaptive. They'll hold up to high pressure and heavy loads. They'll take a beating without getting bent out of shape. They can be cut, melted, glued, drilled, pressurized, depressurized, and more. They are also amazingly simple to work with and reconfigure. In fact, one of the many glorious things about working with a five-gallon bucket—especially when converting it into something exceedingly useful—is that you won't need many specialized tools. Chances are, you pretty much have everything you need sitting in your toolbox right now. The add-ons used to create the projects in this book can all be sourced from any large home center or well-stocked hardware store (some of these projects even repurpose common household items). You may find that leftovers from other home improvement or homesteading projects work perfectly for a five-gallon

bucket project—especially other PVC materials such as plumbing pipes, tubes, and fixtures. That said, the possibilities are even greater with add-ons made specifically for converting five-gallon buckets, available online from plastics suppliers (see **Resources**, page 144).

As if that weren't enough to sway you, the buckets are so ubiquitous that you probably won't even have to pay for one (see **The Bucket Bargain**, page 9, for more info on finding free or low-cost buckets). Even if you do need to purchase a new one, you'll end up shelling out less than you would drop on a morning latte. Quite the bargain!

This is especially true considering the fantastic number of potential uses. The impoverished populations of many developing countries have long understood this. You'll see beat-up, timeworn versions pressed into service for everything from rooftop favela gardens to rural water purifiers to outdoor showers.

The wealth of projects collected in this book includes inventions that range from useful to just plain fun, and from as simple as it gets to fairly involved. These can serve apartment dwellers (**Small-Room Air Conditioner**, page 12); homeowners (**Cyclone Dust Collector**, page 34) and their children (**Air Cannon**, page 94); small farmers (**Chicken Feeder**, page 72); hobbyists (**Post-Mounted Birdhouse**, page 137); and even back-to-the-land homesteaders and survivalists (**Camping and Composting Toilet**, page 48). If you don't see anything that catches your fancy in the list of projects, you must be that rare and remarkable person who already has everything.

Each project also includes at-a-glance indicators describing how much time, money, and expertise you'll need to execute the instructions. These should help you plan accordingly so that no five-gallon-bucket project is left half-finished.

Look through these projects and decide on the ones that best serve your needs or tickle your fancy, or use any of these projects as a jumping-off point for your own one-of-a-kind creations or adaptations. One way or another, you're sure to make something useful, fun, ingenious, and time- and money-saving. Imagine how happy you'll be drinking your own clean, clear water made with your five-gallon-bucket **Water Filter** (page 37). Or how impressed your campground friends will be when they see you using a homemade **Boat Anchor** (page 103) or **Trotline Buoy** (page 101). The rewards and benefits are as plentiful and varied as the projects themselves.

Along with those tangible compensations, if you're reusing a bucket that saw a prior life as a big sauerkraut container or a self-contained supply of wall compound, you'll be saving a bucket from the landfill. The vast majority of plastic buckets are not biodegradable, so the environment thanks you.

CHOOSING YOUR BUCKET

Depending on which project you choose, you'll quickly find that not all five-gallon buckets are alike. The vast majority—but not all—are made of high-density polyethylene (HDPE). This plastic holds up under high temperatures and releases very low levels of contaminants. Low-density polyethylene containers are flimsier and will not tolerate higher temperatures or high-temperature contents. All the projects in this book should be executed using HDPE buckets.

Many of the projects in the pages that follow, and probably many that you can create out of your own imagination, involve consumables of one form or another. Any food or drink requires that the plastic it touches be food grade. Non-food-grade plastic buckets may contain harmful compounds that can leach out of the plastic and into whatever is kept in the bucket— especially if the foodstuff is acidic. This is very important to keep in mind for projects such as the **Water Filter** (page 37).

If you're unsure of what the bucket contained before you got ahold of it, the bucket itself can probably tell you a lot about the original contents. Obviously, a label listing food contents, such as beans, frosting, or salad dressing, indicates food-safe plastic. Symbols on the label also offer clues. A snowflake means that the bucket (and its original contents) can be refrigerated or frozen. A wave symbol or dishes in water mean microwave safe and dishwasher safe, respectively—all signs that the bucket contained food. The manufacturer listed on any label may be a giveaway as well. Do a web search for the manufacturer's name or any code on the bucket, and you'll likely turn up what that manufacturer produces and puts into its buckets.

Lastly, you can look on the bottom in the recycling triangle that is stamped on most buckets. A 1, 2, 4, or 5 inside that triangle tells you that the plastic out of which the bucket has been manufactured is probably safe for food. In most cases, a 7 and the term *bioplastic* mean the bucket is food safe as well.

THE BUCKET BARGAIN

The best price for anything is free. Finding free five-gallon buckets is easy when you consider the many ways this exceptional resource is used. The trick is to hit up businesses that receive raw materials in the buckets but then have no reason to keep them.

• **RESTAURANTS, BAKERIES, AND OTHER FOOD RETAILERS.** Any buckets you recover from these institutions are going to be food grade and are ideal for use in consumable projects. Of course, these buckets are also great for other projects. Make sure you clean the buckets thoroughly though.

• **BUILDING CONTRACTORS.** Large renovation and building contractors often tear through five-gallon buckets of wall compound and regularly toss them right into an on-site dumpster. You'll need to thoroughly clean these buckets before use, but they are some of the tougher buckets you'll find. However, do not go onto a construction or private job site without permission. If the dumpster is on public property waiting to be emptied, it's fair game. That said, it's always nice to ask before taking.

• **SUPERMARKETS.** Grocery stores order many different materials that are delivered in five-gallon buckets. Check the information under **Choosing Your Bucket** on page 7 to determine if buckets scavenged from a supermarket are food safe. You'll usually find the free buckets stacked near a dumpster. Some supermarkets recycle their plastic buckets, so, as always, it's a good idea to seek out the store manager and ask if the buckets are there for the taking.

• **SCHOOLS.** Public school cafeterias and janitorial departments order both edibles and cleaning products in five-gallon buckets. Look for identifying labels and never assume a bucket rescued from a school dumpster is food safe.

• **CAR WASHES AND GAS STATIONS.** Everything from industrial soap to grease can be delivered in five-gallon buckets, and most operations are happy to have the buckets removed. Here again, ask the manager or owner if it's okay to take the buckets.

Sometimes, picking up a perfectly usable five-gallon bucket is just a matter of keeping your eyes peeled. Depending on how sophisticated your local recycling program is, the sanitation department or municipal dump may have buckets on hand that they would be more than happy to see reused. Often, it's just a matter of asking the right person the right questions!

HANDY HOME ADDITIONS

THE BENEFITS OF THE FIVE-GALLON BUCKET BEGIN AT HOME. HOME COMFORTS AND CONVENIENCES ARE SOME OF THE EASIEST PROJECTS TO MAKE USING A STURDY BUCKET, BUT THEY ALSO ADD THE BIGGEST BANG TO YOUR DAY-TO-DAY LIFE. AND IT'S JUST PLAIN FUN TO SEE YOUR FIVE-GALLON CREATION ON A REGULAR BASIS.

These projects can be roughly divided into those that replace home appliances—like an air conditioner—and simpler innovations that replace basic structures you would otherwise have to purchase. Those that stand in for the electrical devices that make our lives easier usually provide the same benefits without the energy consumption. Some can even be wired to a small solar panel. The more basic projects offer alternatives to all those little conveniences, like shelving and shoe racks, which make life easier but to which we rarely give much thought. These are great ways to increase your home's usability and comfort without running out to the store.

Regardless of the final purpose, these projects are all useful and fairly ingenious, and make a minimal impact on the environment. They also bring a dose of fun—and perhaps a conversation piece or two—to the home. What more could you ask from a home addition?

Small-Room Air Conditioner

An air conditioner is a simple luxury that can improve life immensely, especially when it's used to cool a room where you or anyone else spends a lot of time. Add air conditioning to a workshop or home office, and you'll not only be happier and more comfortable, you'll be more productive as well. Lower the temperature in a bedroom on hot August nights, and you guarantee sound, restful sleep. Guests will certainly appreciate the gift of a cool bedroom to settle into during a visit.

Trouble is, even small window air conditioners can be mighty tough on the wallet. They aren't exactly packed with environmental friendliness either. The answer, as always, lies in a five-gallon bucket. Turn this simple container into a basic room-cooling unit that draws warm air in and blows cold air out. You can use this innovative bucket adaptation in any small space to chill a room for the price of a couple iced coffees.

Even though it does the job of a complicated air-conditioning unit, it does it in a simple way. The trick is to use the basic heat transfer physics of ice to your advantage. Instead of an energy-sucking compressor, the unit uses a basic desk fan. The fan draws warm air into the bucket, ice in the unit grabs ahold of the heat and gives off cold, and cold air is forced out of the PVC outlets. Easy peasy, all thanks to basic science.

Once you've built your own air conditioner, you'll want to ensure that it cools your space as effectively as possible. That means positioning the unit where it will have the greatest impact. Keep in mind that cool air is heavier than warm air and will tend to fall, while warm air rises. That means that the best location for a unit like this is up high. You can put it on a shelf, desk, table, or stepladder. Just make sure it's secure and stable. Orient the air conditioner so that the outlets blow wherever you want the air. You should also keep the unit out of direct sunlight, to prevent the ice from melting quicker than it normally would. Under average conditions, with the inside cavity filled with one to two gallons of ice, you can expect the unit to cool for four to five hours.

TOOLS:

Sharpie

Cordless drill and bits

1⅞" hole saw

Jigsaw or hacksaw

Keyhole saw

MATERIALS:

5-gal. bucket with plain lid

1½" PVC pipe, Schedule 40 (12" section)

80-grit sandpaper

Styrofoam 5-gal. companion liner

PVC pipe cement

Small plastic table fan (with cage lip and a base that can be detached)

Silicone sealant

1-gal. resealable plastic freezer bags

Bags of dried beans (or other weight)

HOW YOU MAKE IT

1. Mark 3 holes about 5" down from the top of the bucket (or just below the lowest ridge on the bucket—below the mounting blocks for the handle). Use the end of the outlet PVC pipe as a template to trace the holes, tracing around it with the Sharpie. Drill holes at the marked locations using the hole saw. Use sandpaper to smooth the edges of the holes, and then dry fit the pipe to ensure it fits snugly in the holes.

2. Slip the Styrofoam liner into the bucket and make sure it is secured all the way down inside (the liner can be somewhat flimsy, so don't let it flex inward during fabrication). Hold the liner tightly in place, keeping your hands and fingers safely away from the locations of the vent holes. Use the hole saw to drill through the three holes in the bucket to cut corresponding holes in the liner.

3. Cut the PVC pipe into 3 4" sections using the jigsaw or hacksaw. Dry fit the pipe sections into the holes you've drilled in the buckets. Adjust the fit as necessary, and then apply PVC cement around the outside of each section, along the edge that will be inserted into the bucket. Slide the first pipe into place, so that only about ¼" of the pipe projects into the bucket through the Styrofoam liner. Repeat with the remaining pipe sections and let them dry.

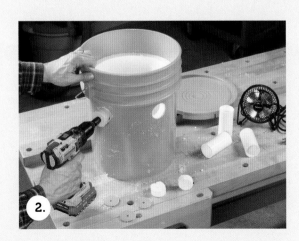

2.

3.

5A.

5B.

4. Remove any stand or bracket from the fan. Center the fan body, face-down, on top of the bucket's lid. Use the Sharpie to trace around the outside diameter of the fan's screen flange. Remove the fan and measure the width of the flange. Mark a second circle, inside the first, representing the inside diameter of the fan's flange (the fan will rest on its flange, when in place over the hole in the lid).

5. Drill an access hole and cut along the inside marked circle with a hacksaw or keyhole saw (A). Work slowly and try to follow the marked circle as closely as possible. Dry fit the fan into the hole, and make adjustments as necessary (B). Once the fit is snug, lay a bead of silicone sealant along the front of the fan flange and press it into place on the lid. Apply a weight, such as a few bags of dried beans, on top of the fan to press it against the lid until the sealant sets.

6. Fill 2 or 3 freezer bags with water and freeze them. (You can place them in a small pail or other mold to create a shape that will fit best into the bucket.) Once the bags are frozen, place them inside the bucket, making sure that none of the vent holes are blocked. Snap the lid onto the bucket, checking that the seal is tight all the way around the lid.

7. Plug in the fan and test the unit. Find the ideal location for the air conditioner, and orient the unit so that the outlets provide a stream of cool air exactly where you need it most. It's a good idea to time how long the air conditioner cools before the ice melts. You can then set a simple timer to alert you when the ice needs to be changed.

PROJECT OPTIONS

There are a number of alternatives to the materials used in this project. Adapt the basic idea here to what you have on hand, recycled materials, or what you can purchase at the lowest cost locally.

• **INSULATION.** If you can't find the Styrofoam liner used in this project (sold through home improvement stores as a "companion cooler"), you can line the bucket with double-sided roll insulation, strips of Styrofoam board insulation, or another insulating product (you can even cut down a beat-up Styrofoam cooler and use the pieces for insulation). The whole idea is to slow the conduction of the cold inside the bucket—and the heat outside of it—through the sides. The colder the inside of the bucket is kept, the slower the ice melts. For the same reason, no matter what material you use, you'll want to be sure to insulate the bottom of the bucket as well as the sides.

• **FAN.** The fan used in this project is an inexpensive desk fan, originally mounted in a stand with a U-shaped bracket. The construction is simple and easy to disassemble. You can use any small fan, as long as the motor is integral (completely inside or part of the wire cage surrounding the blades), and there is a flange of some sort to allow mounting. Although you can use a metal fan, this project uses a plastic one because not only is it cheaper, but the bond between the fan flange and the bucket lid is extremely solid with silicone sealant. If you use metal or another material, you may need to use a different adhesive. You can also consider a USB fan if you'd prefer to power the unit through a computer's USB port.

• **OUTLETS.** The straight outlets in this project can be easily changed to direct cool air upward or off to the side (shown above). All it takes is a PVC plumbing elbow in a mating size to the 1½" PVC pipe. Attach the elbow with PVC cement. If you want to try out a direction, duct tape the elbow in place and use the cement only when you're sure the elbow is blowing air exactly where you want it.

• **ICE.** The core of this unit is ice. Capturing the ice in a separate container, such as resealable bags, saves cleanup time and makes filling or emptying the unit easy. However, you don't have to use the bags specified for the project. Although they are handy because you can freeze them in any shape that works for your bucket, you can also use plastic milk jugs, plastic liter bottles, or another plastic container that you want to recycle. The trick is to fill the interior of the bucket with as much ice as possible without blocking the outlets. The more ice, the longer the cooling will last without a refill.

Cable and Cord Organizer

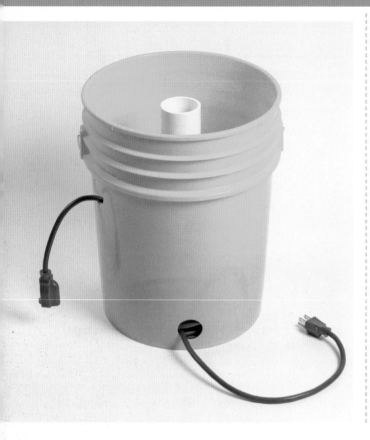

Cables for home computers, peripherals, stereos, and more can lead to annoying visual clutter that ruins the beauty of any room. The chaos of a cable nest also makes sorting out entertainment-center or computer problems a much more difficult task and complicates the process of moving those home electronics to a new room or a new home.

Power cables can be an even worse problem. An extension cord snaking underfoot represents an all-too-real tripping hazard. Unprotected, these cables are at risk of damage that can present other safety issues. Ultimately, controlling cable and cord clutter is just good sense.

A bucket is ideal for organizing these serpentine necessities. The shape allows for the cables or cords to be stored in neat and tidy coils, and a five-gallon bucket can hold a wealth of cables. It can even be used for multiple cables to different electronic components or computer peripherals.

WHAT YOU'LL NEED

Time: **15 minutes** | Difficulty: **Easy** | Expense: **$**

TOOLS:

Cordless drill and bits

80-grit sandpaper

MATERIALS:

5-gal. bucket with lid

4" PVC slip cap

4" PVC pipe (12" section)

(3) ⅜×¾" Phillips-head crown bolts and nuts

PVC primer

PVC cement

HOW YOU MAKE IT

1. Note the height at which you want the cord to enter the bucket and the height at which the other end will exit it. Transfer these measurements to the outside of the bucket and drill holes large enough for the ends of the cable or cord to pass through. Sand the inner surface of these holes smooth.

2. Center the PVC cap upside down inside the bucket on the bottom. Use a ³⁄₈" bit to drill three holes in a triangular pattern inside the cap and through the bottom of the bucket (drill into a scrap piece placed under the bucket).

3. Attach the cap to the bucket with the three crown bolts, hand-tightening the nuts. Prime the inside of the cap rim and the outside of one end of the PVC pipe. Let the primer dry, then coat the inside cap rim and pipe end with PVC cement and slide the pipe into the cap.

4. Once the PVC cement has dried, coil the cable or cord you want to store around the PVC pipe in the bucket. Pull each end of the cable or cord out of the respective holes. Put the lid on the bucket and connect each end of the cable or cord to the outlet or component.

1.

3.

4.

PROJECT OPTIONS

You can stack more than one cord or cable in this manager—just drill access holes at the approximate level of each cable or cord. If you want to separate the cords, you can install the plastic shelf described in the **Shelving** project, page 22. Do not attach the shelf with screws; just stack it on top of the lower cable. That will allow you easy access to the lower cables, should you need it.

A big part of a truly sustainable lifestyle is minimizing the impact of the waste we produce on the environment. Recycling and upcycling are ideal, but some things just have to be thrown out. That doesn't mean that they have to take up a lot of room in a landfill. This trash compactor reduces waste efficiently, without the need to use the energy (or incur the significant expense) of a traditional motorized trash compactor.

Although you can use this convenience in your house to crush solid waste, it's also great for areas such as a summer cabin or long-term campsite that might not have regular trash pickups or local access to a dump. Compact your waste, and it can be far easier to take out with you when you head home. The compactor can also be used to crush yard waste or even kitchen leftovers before they're added to a composting bin or pile.

The device couldn't be simpler. It's basically a weighted bucket that sits atop the waste in another bucket. You can use the brick weights we recommend here, cinder blocks, or metal free weights, which can be found at flea markets, in local newspaper classifieds, or through Craigslist for cheap. The recommended weight is at least sixty pounds, the amount needed to efficiently crush solid waste (of course, the more weight, the better the crushing). If you prefer, you can fill the top bucket with water, which will translate to about forty pounds of crushing weight. The bucket will be easier to lift but slightly less effective in doing its job. You can also use rocks, lead fishing weights, or other similarly heavy items. In any case, make sure you or someone else can lift the weighted bucket. As with all projects, it can be mighty useful to have a helper.

The project also includes an access hole cut in the bottom of the waste bucket to make it easier to push out the puck of compressed waste. You can certainly skip this if you plan on just overturning the bucket to empty it. However, different materials are more or less prone to getting stuck in the bucket. Cutting the access hole is simple and highly recommended.

WHAT YOU'LL NEED

Time: **15 minutes** | Difficulty: **Easy** | Expense: **$**

TOOLS:

Sharpie

Cordless drill and bits

Frameless hacksaw

MATERIALS:

(3) 5-gal. buckets

80-grit sandpaper

Bucket lid

60-lb. free weights, bricks, or equivalent—more is better

HOW YOU MAKE IT

1. Draw a circle with the Sharpie, 5" in diameter, centered on the bottom of one of the buckets, which will become your waste bucket. Cut out the circle by drilling an access hole and using the hacksaw or a drywall saw. It does not need to be exact. Sand the cut edges of the circle smooth with the sandpaper.

2. Turn the second bucket upside down. Drill an access hole and use hacksaw to cut the bottom out of the bucket, sawing right inside the outer rim of the bottom. Here again, you don't need to be precise.

3. Slide the cut bottom down inside the waste bucket, creating a false bottom. Sand the edges as necessary to make it fit so that it can easily be pushed out of the bucket.

4. Stack the weights inside the third bucket and put the lid on the bucket. Put waste into the waste bucket and sit this weighted bucket on top of the waste to compact it.

TRASH COMPACTING GUIDELINES

Follow a few simple rules to ensure your homemade trash compactor works as it should.

• Lift the top, weighted section with your knees. Sixty pounds or more can be surprisingly heavy. If possible, use a helper when lifting the top section on or off the bottom section.

• Remove recyclables. This is especially important where glass is concerned. Never put glass in the compactor because it can shatter and shards could bounce out of the bottom container.

• Drill drain holes in the bottom, if you're compacting for a composting pile. This will limit messiness when pushing the compacted material out of the bottom bucket.

Shoe Rack

These days, "no shoes in the house" has become a common rule in a great many homes. Not only does a shoeless interior mean less dirt, mud, and muck tracked across carpets and wood floors, it's also a matter of home hygiene. Studies show that removing shoes can improve indoor air quality by keeping allergens and other particulates out of rugs and carpets, where they might otherwise exacerbate allergies, asthma, and related conditions.

If you're going to follow the trend, you'll need a handy place for shoes right by the entryway. Because they are often the most common entrance, mudrooms or informal side or back entrances often see more foot traffic (and dirtier gardening footwear). That's where this heavy-duty rack comes in. Positioned by a mudroom door, or under any entryway bench, it provides a convenient place to keep shoes and boots.

The truth is, this home accessory is more function than form—it's probably not attractive enough to sit in a well-appointed entryway or formal foyer. But it really comes into its own in a high-traffic area inside a busy back door.

With an open-weave metal top surface, this rack is ideally suited to sit on top of a plastic tray or other portable, waterproof, and washable surface. A tray meant for underneath a dish-draining rack can be an excellent option, as can a large sheet pan that is too battered for kitchen duty. The rack itself is easily cleaned whenever it becomes too dirty. Just drag it outside and give a good blast from the hose. Let it dry and it will be good as new. You can also spruce it up by painting the bucket a handsome color, and you can even paint the steel top surface a contrasting color. Or paint the whole rack bright white or black to blend right in with your mudroom decor.

Whatever the look, this will serve as perhaps the most durable shoe rack you've ever had. As a bonus, it's super easy to fabricate and assemble, needing no fasteners or modifications other than cutting. To make things even easier, the instructions here include different methods for making the cuts. Choose whichever suits the tools you have or the technique with which you're more comfortable.

TOOLS:

Ratcheting tie-down strap

Sharpie

Measuring tape

Straightedge

Cordless drill and bits

Jigsaw

Metal-cutting jigsaw blade

C-clamps

Hot glue gun and glue

Hacksaw (optional)

Cordless angle grinder (optional)

MATERIALS:

5-gal. bucket with lid

60-grit sandpaper

24x12x¼" expanded metal sheet

HOW YOU MAKE IT

1. Fasten the lid on the bucket, and lay the bucket on its side on a work surface. Secure it with tie-down strap to keep it stable.

2. Draw a cut line straight across the center of the bottom, parallel to the work surface. Draw another line, parallel to, and approximately 3½" down from, the top line. This will be the bottom edge of the shoe rack.

3. Measure the distance from the point where the bucket contacts the work surface up to the bottom line. Mark this measurement on the lid, and draw a bottom cut line on the lid. Measure up from this line 3½". Draw a top cut line parallel to the bottom line.

4. Use the straightedge to extend top and bottom cut lines along the bucket's sides.

5. Drill a pilot hole on the top cut line. Use this to start the jigsaw cut. Repeat with the bottom line, sawing along the sides and finishing with the lid. Make the cuts with a hacksaw if you have problems sawing through the lid with the jigsaw. Sand all the cut lines smooth.

6. Measure the dimensions of the cut bucket top opening (the width will be slightly different end to end, due to the bucket's taper). Subtract ½" and use the Sharpie to transfer these dimensions onto the expanded metal sheet.

7. Clamp the expanded metal sheet to a work surface with the edge sticking out so that the cut line is unobstructed. Cut along one marked line using the jigsaw with the metal-cutting blade or an angle grinder. Rotate the sheet and repeat for each of the remaining three lines.

8. Position the cut expanded metal sheet over the top bucket opening and carefully press it down. Press the edges down evenly until the sheet is stuck in the opening. Dab hot glue at several of the contact points—the more, the better. Sand any high points on the bottom edge as needed to ensure the rack sits flat.

2.

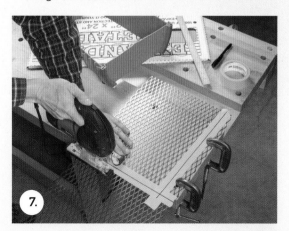

7.

This much is true in a house, garage, and workshop alike: there is no such thing as too much storage. Shelving is some of the best storage because it allows you to keep everything organized in one easy-to-access place and in plain sight. Shelves are also space-efficient storage options. There are lots of shelving options, but a five-gallon bucket makes for a wonderful, self-contained shelving unit. Not only is a bucket the right size for a manageable shelf, but individual bucket shelves can be combined to create a shelf tower or even a wall of shelves as needed.

Turning buckets into shelves is a simple affair. The most challenging part of this particular project is cutting the plywood circle. Even though that shouldn't tax your skills no matter what those skills may be, if it seems like a little bit more than you want to tackle, you can opt for the quick-and-simple solution of cutting a circle out of the bottom or lid of a sacrificial bucket. Cut the circle just a bit smaller than the top diameter and then push it down into the finished unit where the pressure of the sides with the lid on will keep the shelf in place (shave off the edges with a utility knife to make small adjustments). Just keep in mind that an unsecured plastic shelf won't support the heavy weight a screwed-in plywood shelf would. If you want to add stability, screw simple mounts to the inside of the bucket to hold the plastic circle in place.

The shelving can be used wherever there is a need, although it may not be entirely suitable for highly visible areas. Stack three of these units in a kitchen alcove for a handy pantry that is easy to clean. Or screw one to the wall in a mudroom for gloves and other foul-weather gear. The handle adds another dimension to this shelving unit. It allows the shelf to be hung from rafters in a garage or workshop, to keep tools and supplies at eye level, and makes it easy to move the unit at a whim. Just make sure your rafters can support the weight of whatever it is you want to store in the bucket.

WHAT YOU'LL NEED

TOOLS:

Sharpie or carpenter's pencil

Straightedge

Frameless hacksaw

Measuring tape

Carpenter's compass

Torpedo level

Cordless drill and bits

MATERIALS:

5-gal. bucket with lid

80-grit sandpaper

1" plywood scrap, at least 12×12"

(3) 1" Phillips roundhead wood screws

HOW YOU MAKE IT

1. With the lid removed, mark cut lines from top to bottom on the bucket as shown in the photo, running up each side on the inside of the handle mounts so that the shelf opening takes up slightly less than half the diameter of the bucket. With the lid removed, use the straightedge to mark the lines to points just above the bottom. Mark a horizontal line from the bottom of one vertical line to the other.

2. Use the hacksaw to make the cuts, starting at the top. Cut down one side, across the bottom, and up the other side. Sand the cut edges smooth.

3. Carefully measure the width of the bucket interior at the point where you want to position the shelf. Transfer this measurement to a small sheet of plywood, near one edge. Mark the center point. Position the compass point on the center point and draw a circle.

4. Cut in from the edge of the plywood with the jigsaw and carefully cut the circle out of the plywood sheet. Sand the edge smooth.

5. Position the shelf in the bucket (with bucket lid fastened on top). Use the torpedo level to ensure the shelf is level. Mark the outside of the bucket for the screws that will hold the shelf in place—one on either side and one on the back.

6. Drill pilot holes for the screws through the bucket and into the edge of the shelf. Secure the shelf with the roundhead screws.

2.

4.

6.

A FIVE-GALLON SHELF TOWER OR WALL

Individual five-gallon-bucket shelving units can be combined in different ways to create columns or even whole walls of storage. This could provide all the storage you might need along one wall of a garage or workshop, and the pattern of the shelving can be adapted to suit the space available, what you need to store, or your own preferences. No matter how you organize the shelves, you should follow some basic safety rules to ensure that they are as stable and secure as possible.

Stack columns of bucket shelves so that each successive bucket stands on the cover of the bucket below. The trick with this design is to ensure that buckets on top don't overload the lids on which they are standing. Store heavier materials on the bottom shelves, and position increasingly lighter materials as you fill the shelves above. Overall, this is a good shelving structure when you need to organize and store lighter materials such as tubing, hoses, plastic fittings, or out-of-season clothing.

For a sturdier shelf structure—one that can support heavier materials on all the levels of shelves—create a running bond pattern stack. Center a bucket over where the two edges of the buckets below it meet. This creates a modified pyramid and ensures that the load is transferred from one bucket down through the sturdy walls of the buckets below it. In this formation, you can stack heavier items in upper shelves as well as lower shelves, and the shelving unit can in general support a lot more weight. This would be a good design for storing tools, gardening supplies, or family-size bags of food such as rice or flour.

Weight is just one issue. You also want to make sure that the shelving design you choose is stable. Regardless of whether you're building a shelving tower or a pyramid, screw the buckets in the second row and above to the wall. Either drive the screw directly into a stud, or use a wall anchor. One screw per bucket should suffice.

Lastly, to make the shelves more attractive and more suitable for use in highly visible areas of the house, see **Painting Your Bucket**, page 56.

Package Mailbox

This unit may not necessarily replace your primary mailbox. It's a bit large just for mail, and the US Postal Service maintains regulations on the size, shape, and location of an official home mailbox. These regulations are subject to the discretion of the local postmaster, so that's the place to start if you think you'd like to use this mailbox as *the* mailbox.

Even if this isn't your primary mailbox, this structure is great for boxes delivered to your address, especially those that are larger than would comfortably fit in a mailbox. In this age of prolific online shopping and store-to-door delivery, there's something to be said for a convenient drop box that will keep your purchases out of the elements until you get home. The large capacity means you can buy something like a winter coat or a new lamp, and the box it comes in will likely fit comfortably inside the mailbox.

This is also a way to secure packages that will be sitting for some time before you can retrieve them. If you've bought and paid for something valuable, or even something not so valuable, it's no fun when it walks away from your doorstep. It can sometimes be hard to prove that the package was even delivered. However, if you're not concerned about security, you can skip the step involving drilling the lock holes.

As a bonus, the base can double as a mini flower garden, providing beauty along with the convenience of a large-package shelter. If you'd prefer not to deal with tending plants, you can certainly fill the top third of the foundation bucket with decorative stone, thick coiled rope for a nautical theme, or any other decoration that suits your fancy (try colored lights for a holiday theme!). You can paint or stencil the base and mailbox buckets for an even nicer look.

WHAT YOU'LL NEED

Time: **1½ hours** | Difficulty: **Moderate** | Expense: **$$$**

TOOLS:

Measuring tape

Sharpie

Circular saw

Torpedo level

Keyhole saw, crosscut saw, frameless hacksaw, or jigsaw

Cordless drill and bits

Straightedge

C-clamps

MATERIALS:

4×4×42" pressure-treated pine post

(2) 5-gal. buckets, one with lid

Bag of quick-setting concrete mix

Piece of scrap or stick

Brads

(3) 3" flathead wood screws

(2) 2½" stainless steel butt hinges and screws

Bag of potting soil (optional)

Plants (optional)

HOW YOU MAKE IT

1. With lid removed, stand the bucket on a solid, level surface, such as a concrete slab. Hold the post upright in the bucket, centered on the bottom of the bucket. Secure the post in place by tacking 2 long scrap pieces of 2× wood along two sides of the post with the brads (you can substitute any sturdy scrap pieces that are long enough to bridge the mouth of the bucket—use longer nails as necessary). Then pour in quick-setting concrete mix, filling the bottom third of the bucket around the base of the post.

2. Slowly add water, tamping the concrete with the stick or scrap piece of wood as you pour the water in. Continue adding water and agitating the concrete until the concrete is evenly wet and the consistency of cookie dough. Rap the sides of the bucket to release any air bubbles in the concrete.

3. Use the level to check the post for plumb. Be sure to check two perpendicular faces. Check for plumb again, adjust as necessary, and then allow the cement to dry.

4. Remove the braces from the post. Drill four ⅛" holes equidistant around the bucket, just above the top level of the concrete (for drainage). Fill the bucket with potting soil (or use gravel or decorative stone if you don't intend to plant the bucket).

5. Remove the lid from the second bucket and lay the bucket on its side, centered front to back, on top of the post. Use a nail to mark three drill holes inside the bucket, in a triangle over the top of the post. Drill the three holes through the bucket and down into the post. Screw the bucket to the post with three flathead wood screws.

2.

6. Measure 2" in toward the center of the lid from the edge, and use the straightedge to mark a straight line across the lid running through the mark. Carefully mark the rim and edges of the lid to continue the line at each end.

7. Clamp the lid to a work surface with the cut line side hanging over the edge and free of any obstruction (or secure the lid in a vise). Use a saw to cut along the marked line, and cut the lid into two pieces.

8. Clamp the two pieces of the lid together as tightly as possible, securing them so that the cut line is over the work surface. Position the butt hinges so that they are equidistant from each side and each other. Screw the hinge flanges on either side of the cut to the lid on each side.

9. Snap the top section of the lid onto the mailbox bucket. You can push the bottom part to snap it close, or leave it unsecured so that it is easier to open. Paint the word *Packages* or *Mail* on the side and lid with a stencil, if necessary. Plant flowers or other plants in the soil if you desire. Position the box where you want it and check for level.

5.

7.

PROJECT OPTIONS

Turn your package drop box into a lock box with a drill and a padlock. Mark a location on the bottom half of the mailbox, right under the ridge where the lid snaps into place on the bucket. Now mark a corresponding hole at the edge of the lid, in line with the first hole. Use a three-quarter-inch bit to drill holes at both marked locations, and use a padlock with a long tang to lock the drop box. You can leave the box unlocked for deliveries with a note for the delivery person to lock it—and never worry about a valuable package going missing again.

8.

Cat Litter Box

Even cat lovers know that a well-designed litter box is worth its weight in gold. If you have a feline friend of the indoor-cat variety, there's no getting around the fact that you need a cat litter box. The trick is to choose a box that is easy to clean, confines unpleasant smells, and takes up as little space as possible. You can find a number of models at retail, from simple rubber trays that are nothing more than containers for litter to covered bathrooms that include holders for air fresheners. But you (and your favorite kitty) will be just as well served by a thoughtfully designed, homemade five-gallon-bucket model.

The best litter box affords the animal a modicum of privacy. Cats are discreet, and they like to be out of view when doing their business. The design for this project is enclosed, which serves two purposes: the shy kitty is kept concealed during his or her "me" time, and any odors are better contained than they would be with an open, tray-type design. An enclosed structure also means that the cat can't scatter litter all over the floor, or accidentally spray a nearby wall or fixture.

The number one selling point for this particular design, though, is cleanability. The half lid on the front can simply be removed to pour out dirty used litter when scoop cleaning won't do the trick, and it facilitates easy litter replacement. Not only that, but when emptied, the bucket can be sprayed out with a little dish soap and a garden hose, making it fresh as new.

Using the handy wood cradle, the box can be placed just about anywhere you might want it to go (and you want the cats to go). That said, left plain, the box will look best in a utility area such as a mud room, large guest bathroom, or even a closet. If you want to place it in a frequently used bathroom or other high-profile area, you might consider painting the bucket and the cradle. Keep in mind though, if you use a white bucket, light will bleed through and ensure that your cat doesn't have to go in the dark. You can also tie a chair pad to the bucket to create a pet bench on top.

WHAT YOU'LL NEED

TOOLS:

Measuring tape

Sharpie

Straightedge

Vise or clamp

Jigsaw or frameless hacksaw

Cordless drill and bits

MATERIALS:

5-gal. bucket with lid a

2"×2"×5' pine (or similar)

(12) 3" flathead wood screws

½×11" section of polyethylene self-seal pipe wrap insulation

Bag of kitty litter

Self-adhesive furniture pads (optional)

HOW YOU MAKE IT

1. Remove the lid from the bucket and measure and mark a centerline across the lid. Mark a parallel line 1" below this line— that will be the actual cut line. Clamp the lid in a vise or to a work surface, and use the jigsaw or hacksaw to cut the lid in half, along the cut line as a guide.

2. Drill several vent holes in a random pattern along one side of the bucket. This will be the top of the bucket when the bucket is lying in the cradle.

3. Build the cradle. Cut 2 sections of 2×2 12½" long, 1 section 7" long, and 4 sections 6" long. Assemble the cradle as shown in the illustration, drilling pilot holes and driving two 3" screws through each joint.

4. Tilt the bucket away from the holes on the side and fill it about ⅓ full of kitty litter. Snap the lid onto the top on the side opposite the holes. Lay the bucket in the cradle and shake very lightly to evenly distribute the litter across the bottom of the bucket. Add litter as necessary to fill the base of the bucket up to the level of the lid's cut side.

5. If the bucket is not firmly held by the cradle arms, stick self-adhesive felt furniture leg pads to the arms to hold the bucket in place. Cut the pipe insulation to length, line the inside with a bead of silicone sealant, and secure it over the cut edge of the lid. Let the sealant cure before letting your cat use the litter box.

1.

3.

5.

Mousetrap

Mice are bothersome pests that, left to their own devices, can multiply into a serious problem and even an infestation in short order. Not only do they eat or otherwise destroy food, they also contaminate food areas with feces and can even spread illness. They commonly gnaw through wiring and can create serious situations in walls that can be quite expensive to fix.

The secret to making sure a mice problem doesn't get out of hand is to deal with it quickly and completely. You can poison the little critters, but that can lead to as many problems as it solves (decomposing mice in the walls are not a smell anyone wants in their home). That means using the best mousetrap you can find— or in this case, make.

There are many different mousetrap designs on the market, all with their own pros and cons. But really, why not make your own, super effective mousetrap? It's easy to do with a few implements you probably have lying around the house right now. This particular design is also fairly ingenious, and mice will usually not learn to avoid it, as they might with other traps.

The beauty of this design is that it can be used time and time again, unlike a glue trap. And unlike a snap trap or other kill traps, disposing of the mice is easy with a five-gallon-bucket trap; no need to touch the bodies at all! It will be up to you to decide whether you want to kill the mice or catch and release them. Catch and release is a much more humane method (take the mice at least a mile from your home before releasing them into a park or other semiwild area). If you opt to kill them, fill the bucket about one-quarter full with antifreeze; it will kill the pests quickly and will diminish any telltale odor.

Keep in mind that any mousetrap—this one included—needs to be positioned and baited correctly to be effective. Read the guidelines in **Keys to Capturing Mice**, page 31. You should also continue using the trap until you're absolutely certain that there are no mice left in the home.

TOOLS:

Cordless drill and bits

MATERIALS:

Sharpie or pencil

¼" wood or steel dowel (at least 15" long)

5-gal. bucket without lid

Scrap of 1" PVC pipe (6 to 8" long)

1×3" (or similar) scrap, 15 to 18" long

Duct tape

HOW YOU MAKE IT

1. Lay the dowel across the top of bucket and mark the locations of the drill holes about 1" below the lid. Drill ¼" holes for the dowel on each side of the bucket at the marked locations.

2. Slide the dowel in through one hole. Slide the PVC scrap onto the dowel, and push the dowel out through the opposite hole. (You can use a soda can if you don't have a scrap piece of PVC pipe).

3. Duct tape one end of the 1×3" scrap to the lip of the bucket at one end of the dowel, resting the other end on the floor to create a ramp. Slide the PVC pipe section to the ramp end of the dowel and bait the top of the pipe with a generous dollop of peanut butter mixed with sunflower seeds; it should spin freely around the dowel to drop any mouse that tries to get at the food into the bucket. Position the trap where there is evidence of mice, and regularly check the trap. Re-bait as necessary.

KEYS TO CAPTURING MICE

Reusable mousetraps are most effective when you position, bait, and maintain the trap carefully. Here are a few things to keep in mind:

• Position the trap where mice have been active in the last twenty-four hours. (Look for droppings, gnawed wiring or wall surfaces, or chewed food.)

• It's best to position the trap along a wall, away from foot traffic, and not in a corner. Try to establish the vermin's paths of travel (mice often scurry along a wall to get to one part of a room or another), and place the trap in a regularly used path.

• Bait the trap with peanut butter and birdseed to start with. If you're not successful, change the bait. Chocolate and bacon are popular baits as well.

• If you're not using antifreeze inside the bucket, check and empty the trap often or other mice will become wary of it.

Acoustic Speaker Dock

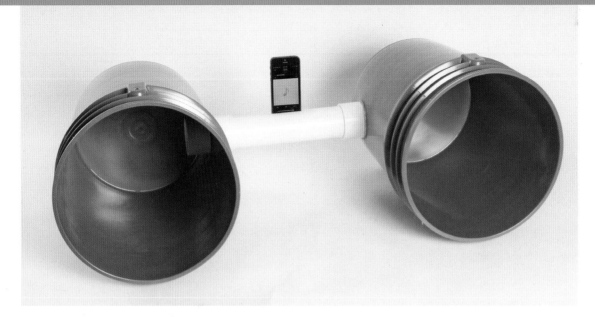

Turning your cell phone or other portable music player into a mini portable stereo is simple these days. You can use one of the many self-contained portable speaker systems on the market, which let you plug and play: you simply connect the music player and presto—big sound. The downside to these systems is that they can run into the hundreds of dollars and most need their own electrical connection to operate (or else they'll burn through batteries at a fast clip). Battery-operated versions can run out of power long before your party's over. These systems can also be a little delicate—you don't want to get a lot of water, sand, or dirt on them or you'll wind up having paid a pretty penny for the equivalent of an electronic paperweight.

Fear not. Even if your budget is tight and your party will be nowhere near an electrical outlet, you can still jam the tunes using your music player. Create a much cheaper, fun, efficient, durable, and portable speaker system alternative with little more than a two five-gallon buckets, a short piece of PVC pipe, and a small bit of your time.

This speaker system is simple as can be. The secret to its success is that it exploits basic audio science. As anyone who has ever lived near a busy freeway can tell you, sound waves travel (sometimes great distances) and bounce—especially off of hard, smooth surfaces. Like the inside of a five-gallon bucket, for instance. Amplify the modest built-in speakers on your music player, and then bounce that sound all around a big echo chamber, and you've got a speaker system that does a lot with a little! It can also take a beating, so you'll never have to worry about it getting wet, or dirty, or bouncing around in the trunk of your car.

You can modify this setup to accommodate whatever music player you own. You can also jazz up your own personal five-gallon-bucket boom box with band stickers, or a Sharpie, and some creativity. Whatever your own personal look may be, you'll control the volume of the system right from the player. Most players mount the speakers on the bottom of the unit; if your music player's speakers are on the top, you'll need to mount the player upside down in this system.

TOOLS:

Vise

Measuring tape

Sharpie

Cordless drill and bits

2⅜" hole saw

MATERIALS:

2" PVC pipe (10" section)

(2) 5-gal. buckets without lids

60-grit sandpaper or a fine rat-tail file

PVC cement

(2) PVC slip couplings

Silicone sealant

HOW YOU MAKE IT

1. Secure the PVC pipe in a vise. Measure along its length and mark the center point. Center the bottom of the music player on the center point and trace around the player's base with the Sharpie.

2. Drill out the player's hole, using a drill bit with a diameter slightly smaller than the width of the mounting hole. Sand or file the cut edges smooth.

3. Measure and mark points 3" up from the bottom of both 5-gal. buckets. Position the end of the PVC so that the bottom edge sits on this mark, and trace around the pipe.

4. Drill out the circles with the hole saw. Dry fit the PVC pipe into one hole. The fit should be very tight. If the pipe end will not go into the hole, widen the hole a little bit at a time with sandpaper or a round file. Repeat with the second bucket.

5. Lay each bucket on its side with the holes facing each other. Stick one end of the pipe through one of the bucket holes (the player's mounting hole should be positioned straight up). Coat the pipe end with PVC cement and slide a PVC coupling onto the pipe, on the inside of the bucket. Repeat with the opposite bucket.

6. When the PVC cement is dry, lay a bead of silicone sealant around the seam where the pipe enters each bucket. Allow the sealant to dry, and then try out your new speakers.

4.

2.

5.

Cyclone Dust Collector

A cyclone dust collector, also known as a cyclonic separator, performs a basic function courtesy of some very cool science. Add inlet and outlet ports to a plain old five-gallon bucket, and you create an empty space for dust to collect. Connect one port on the bucket to the dust collector from a table saw or other power tool and the other port to a basic shop vac, and you create a potential vortex. When the vacuum is turned on, it starts a whirlwind inside the bucket, mimicking the action of a tornado (or, if you will, a cyclone). The effect produces suction that draws the dust from the dust collector, and centrifugal force spins the dust particles outward, so that they hurl against the surface of the bucket and collect at the bottom. The mostly clean air then gets sucked out and into the shop vac.

Cyclonic collectors are used by woodworkers, metal workers, ceramic craftspeople, and even pet groomers. The purpose is always the same: remove contaminants from indoor air. A properly functioning dust collector also saves on vacuum filters, as well as extending the life of the shop vac and making shop cleanup easier. Airborne particles being a major contributor to asthma and other lung conditions, a dust collector can be an essential health-related part of your shop setup—especially if you do a lot of work there.

You can go out and buy a complete turnkey cyclonic system, if you're fond of spending a few hundred dollars. If not, you can build one yourself for the about the cost of shop vac filter, using a few simple parts and less than an hour of your time.

Keep in mind when fabricating this project that an airtight seal is everything. The lid for the bucket you use has to snap onto the bucket securely, and all the connections have to be made carefully to ensure a perfect seal. You may also want to invest in a small dolly to make the collector more portable, or buy four casters and make your own dolly using a scrap sheet of plywood. Lastly, just to be safe, wear a dust mask whenever you empty the collector or clean or change the filter in your shop vac.

TOOLS:

Sharpie

Cordless drill and bits, including 3" hole saw

MATERIALS:

5-gal bucket with lid

(2) sections 3" PVC pipe (4" each)

Silicone sealant

PVC primer

PVC cement

(2) 3" PVC long-turn 90° elbows with slip-fit couplings

(2) rubber no-hub fittings

(4) hose clamps

HOW YOU MAKE IT

1. Mark two holes on either side of center on the bucket lid. Use the PVC pipe end as a template. Drill a pilot hole in each circle and cut out each with the hole saw.

2. Prime the ends of the elbows and the slip fittings with PVC primer and then coat them with PVC cement. Fit the pieces together through the lid, making sure the elbows are on the underside of the lid and are pointed in opposite directions.

3. Lay a bead of silicone sealant around the base of the elbows' flanges and allow to dry before continuing.

4. Snap the lid into place on the bucket. Tighten one end of each of the rubber fittings onto the pipes on top of the lid and connect the other ends to your shop-vac hose and power-tool hose using the hose clamps.

5. Test the system by running the power tool and shop vac. Check for air leaks around the hose clamps and lid. If you find any, tighten the clamps or replace the gasket in the lid channel.

PROJECT OPTIONS

The science of the project is fairly basic and well served by the inlet-outlet port design. However, there are superior designs that may work slightly to significantly better. You may want to consider one of these if the cyclonic separator you've constructed doesn't work up to your expectations for whatever reason.

If you're willing to spend from $25 to $100, you can buy a prefabricated aftermarket cyclonic separator meant specifically to work with a five-gallon bucket. These add-ons are simply attached to the top of the bucket, or fastened to a hole in the bucket's existing lid, depending on the design. There are different designs to choose from, but price is usually a fairly reliable indicator of effectiveness. Whichever you choose, all come with preformed, integral inlet and outlet, making hookups simple and easy. These cut down on the amount of work you'll need to do—at a considerable reduction in the cost of a full-scale cyclonic separator.

A conical separator such as the Dust Deputy produced by Oneida Air Systems (see **Resources**, page 144) creates a highly effective cyclone effect within the unit itself. The five-gallon bucket just becomes a large-capacity cache for the dust and debris processed through the cyclonic separator on top. Common claims for this type of separator boast that ninety-nine percent of dust will be removed prior to making it to the shop vac—greatly increasing the filter life in the vac.

Simple lid replacements such as the Mini Dust Collection Separator from Woodcraft (see **Resources**) are not only simpler than other designs—this design is actually similar to the project described here—they are easy to install as well. Just replace a five-gallon bucket's lid and you're ready to go. The bucket in this case is used to create the cyclonic action as well as collecting the dust and debris. The one-piece construction ensures against air leaks or breakage. This can be a wonderful option for the home woodworker or shop craftsman.

Water Filter

A homemade water filter is an awfully handy addition to your home backup supplies. The filter can be essential in an emergency, such as a flood or earthquake, where the existing water supply may be comprised. But a high-volume water filter is also a lifesaver in the event that a well that serves as a primary source of drinking water for the house becomes contaminated or otherwise unusable. If you're willing to portage it in, this appliance can also serve you well for filtering stream, river, or even lake water that you suspect is not drinkable.

The water filter described here uses three gravity-feed-activated charcoal-and-ceramic filters. You'll find a wide range of these on the market, but the "candle" style (see **Resources**, page 144) used in this project is the most common for homemade filtration systems. The difference between different filters on the market is what they filter out. More expensive models will remove even trace amounts of industrial chemicals and heavy metals. Most of them will remove larger particulates and more common contaminants such as lead.

If you are concerned that the water you're filtering may contain microbes or particulates that your filters don't treat, you can use one of two methods for secondary purification. Either boil the filtered water for five minutes at an active rolling boil, or add eight drops of chlorine bleach per gallon of water, and let the water sit, uncovered, for at least an hour before drinking it. You can also combine these two for even more protection. Either way, thoroughly test the water filter long before you have a need to use it.

TOOLS:

Cordless drill and bits

Groove joint pliers (optional)

MATERIALS:

Sharpie

(2) food-grade, clean 5-gal. buckets with lids and handles

(3) gravity-feed, candle-style water filters

Plastic water dispenser replacement spigot

HOW YOU MAKE IT

1. On the underside of one of the bucket lids, mark three holes in a triangle around the center of the lid. The holes should be separated from each other by about 5".

2. Turn one of the buckets upside down and center the marked lid, upside down, over the bottom of the bucket. Use a 7/16" bit to drill holes at the three marks, through the lid and the bucket bottom. (Or use the drill bit size that corresponds to the nubs on the filters you are using.)

3. Working from inside the bucket, push the mounting post of one filter through a hole in the bottom, and through the corresponding hole in the lid. Tighten the filter in place using the supplied wing nut. Hand tighten the nut until it is very snug. Repeat with the remaining filters. Fill the bucket with about ½ gal. of water and let stand to check for leaks. If you detect any leaks, tighten the mounting nuts until the seal is secure.

4. Mark a hole in the second bucket on the side, 1" up from the bottom, for the spigot. Use the post of the spigot as a template to mark the hole, and then use the appropriate bit to drill the hole. Snug the nut on the spigot to ensure a tight seal (using pliers if necessary), and then fill the bucket with water to just above the spigot and check for leaks.

5. Prime the filters as necessary, according to the manufacturer's instructions. Place the bucket and lid holding the filters on the lower bucket, and snap the lid in place. Fill the top bucket with 5 gal. of water and run this load through the filters and discard. You're now ready to filter contaminated water. The 3 filters should process 1 gal. of water every 15 minutes.

3.

4.

Gravity Cat Feeder

Nobody wants their pets to go hungry. But in the hustle and bustle of modern life, it's easy to forget your furry friends. Making sure that there is food in a dog's dish or cat's bowl is one more chore that can get overlooked. There are also situations, like when you want to spend a weekend away, in which feeding your pets is just not possible. It would be nice not to have to impose on a friend or relative to drop by just to set dinner out for your tabby (or worse yet, incur the expense and hassle of a few nights in a kennel). Now here's a way to ensure that your best furry friend has all the food he or she needs, even when you're not at home.

And, although dogs don't use litter boxes and will still have to be let outside, the feeder

can be used for a dog if you want to avoid having to put out new food for your dog every day.

The feeder works on a simple gravity principle and, given the five-gallon capacity, could go a long time without needing to be refilled. However, it's always wise to periodically check the feeder to ensure that it hasn't attracted unwanted attention; if you have any issues with rodents in your house, this feeder could be an open invitation. The tight seal of the lid should ensure that the food inside the feeder doesn't go stale.

It's important to note that—as useful as this feeder can be—there are certain situations in which a gravity pet feeder isn't ideal. If your cat or dog has food issues such as overeating, or particular health concerns that require a stringent diet, this feeder is obviously not the best option. In those special cases, a preprogrammed portion feeder would be a better idea. Although pricey, they are a good alternative where your animal's health is concerned.

The feeder can also be adapted for homestead barnyard duty. Set up on a support such as cinder blocks, it could feed a goat or two and possibly other small farm animals as well. As always, don't be afraid to use the project as a springboard for your own inventive designs.

WHAT YOU'LL NEED

Time: **30 minutes** | Difficulty: **Easy** | Expense: **$$$**

TOOLS:

Jigsaw or frameless hacksaw

Hot glue gun and glue

Sharpie

Cordless drill and bits, including a 3" spade bit

Straightedge (optional)

Utility knife (optional)

MATERIALS:

(2) clean, food-grade 5-gal. buckets with lids

Large circular rubber or plastic pan (the bucket must be able to sit inside)

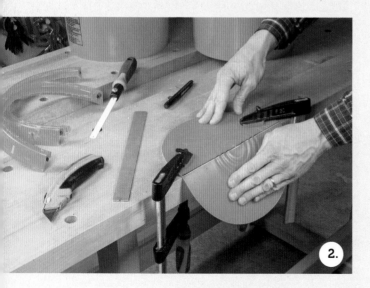

2.

HOW YOU MAKE IT

1. Cut one lid just inside the rim to create a flat circle. Use a jigsaw or hacksaw and follow the curve of the rim—but you do not need to be exact.

2. Lay the circle on the edge of a work surface, so that exactly half of it is sticking out past the edge. Carefully bend the circle down to create a crease across the center of the disk (a very light score along this line, following a straightedge and using the utility knife, may make the bending easier). Wedge the circle down in one of the

buckets, so that one end point of the crease is touching the bottom on one side of the bucket and the other end point of the crease is wedged against the bucket side.

3. At the point of the crease on the lower side, mark the bucket by pushing a nail through to the outside (you can also just eyeball the location). Use this as a reference point and draw a rough "mouse hole" opening on the outside of the bucket around the cut. The hole should be approximately 3" at its highest, curving down to the bottom edge of the bucket on each side to create an opening about 4" wide. Drill a pilot hole and use the jigsaw or frameless hacksaw to cut out the hole. Dab with hot glue to hold this ramp in place, creating a slide for the cat food.

4. Use the spade bit to drill a hole in the center of the second bucket's bottom. Place the bottom bucket in the circular pan so that it sits flat. Slowly slide the top bucket into the bottom bucket. Fill the top bucket, slowly, with dry cat food. When the bottom is full, some food will spill out of the opening, and then the food will begin filling the top bucket. Fill it about halfway and secure the second, uncut lid on top.

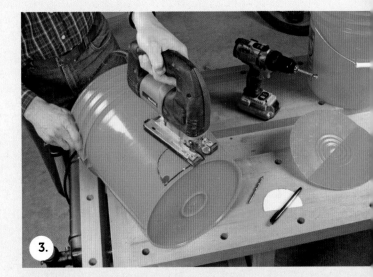

HANDY HOME ADDITIONS

Manual Washing Machine

Eventually, we all need clean clothes. But few people enjoy laundry day. That's why washing machines are such a wonderful modern convenience. Just throw a load (or, in many cases, less than a load) into the unit, turn it on, and presto—clean clothes. The downside is that these magnificent machines are energy and water hogs. Depending on how new and what type your washing machine is, it can use between fifteen and forty-five gallons of water per load!

Ironically, these machines use amazingly sophisticated technology to perform a rudimentary process that hasn't changed much in centuries. The basic principle of that process involves agitating dirty clothes in soapy water for a prescribed period of time. Use warm water for colorfast clothing and white textiles, and cold or warm water for garments with colors that might bleed. Rinse in the same temperature. Basically, that's all there is to it.

Truth is, that process is so simple that it can easily be done with a manual washing machine. The one described here will be highly effective at getting clothes clean without a power source. It's an ideal convenience in the event that your home loses power for a significant period of time, or if you go on a lengthy RV camping trip. Whether you're staying in a cabin, a camper, or a tent, or just want to simplify your homestead, having a manual washing machine at your disposal means you don't have to have quite as many clothes.

The other benefit is exercise. So many people pay for a gym when you can get a decent workout just by saving electricity and water. Even the least fit person in the household will be quite capable of washing a small load of clothes, while burning a tidy sum of calories. You can even operate the washer described here while sitting down if need be, or it can be used on a table or other raised surface so that someone with back problems doesn't have to bend over.

Whatever the case, because the whole idea is environmental friendliness, it's best if you use biodegradable soap to maintain the green cred of this appliance. That way, you can reuse the water in a nonedible part of your garden.

MANUAL WASHING MACHINE TIPS

This is a very simple device, but one that needs to be used correctly if you're going to get your clothes clean with a minimum of effort. The following guidelines will help you do that.

• **USE ONLY BIODEGRADABLE LAUNDRY DETERGENT.** The whole point of this "machine" is to save electricity and water, and it should be used as a green water source.

• **YOU ONLY NEED A LITTLE DETERGENT;** many people use too much laundry detergent. Keep in mind that the capacity of this washing machine is much less than a standard household unit. A tablespoon of detergent should be plenty. Coupled with vigorous agitation, your clothes will be as clean as ever.

• **DISASSEMBLE THE MACHINE AND LET THE PARTS DRY BETWEEN LAUNDRY DAYS.** If you leave the lid on while the machine is wet, mold can grow in the bucket.

TOOLS:

Cordless drill and bits, including 1½" spade bit

MATERIALS:

5-gal. bucket with lid

Cup plunger with 6" or larger head

80-grit sandpaper

Plastic water dispenser replacement spigot

Silicone sealant (optional)

HOW YOU MAKE IT

1. With the lid secured on the bucket, use the spade bit to drill a hole in the center of the lid. Check to see that the plunger handle will slide through the hole with a fair amount of play, allowing it to be moved around inside the bucket. Sand the edges of the hole smooth so that they don't abrade the plunger handle.

2. Drill a ⁷⁄₁₆" hole (or use the appropriate bit size for the spigot you're using) in the side of the bucket, as close to the bottom as possible. Fasten the water dispenser replacement spigot in the hole, tightening it enough to ensure a good seal.

3. Drill 6 to 8 holes ⅜" in diameter in a random pattern on the rubber cup of the plunger. Test the washer by adding a small load of clothes, filling with water and a small amount of detergent, and plunging the load. Move the plunger around to ensure complete agitation. After 10 to 15 minutes, drain the water through the spigot and refill with clean water. Agitate again and drain. Repeat as necessary. If the spigot leaks, empty the bucket, let it dry, and lay a bead of silicone sealant around the outside rim of the spigot. Let the sealant dry, then fill the bucket with water and check that it no longer leaks before using.

Portable Wine Rack

The easy-carry handle and impact-absorbing construction of a five-gallon bucket make for a perfect traveling wine rack that ensures four bottles of your favorite vino will arrive safe and sound no matter where you're going or how rough things get on your way there. Of course, you might be the type of oenophile that prefers to keep your wine stash safe and sound at home. In that case, you can craft multiples of this rack and stack them to store your favorite vintages in the basement, garage, or other cool, dark place.

Foam insulation board in the bottom of the bucket protects the bottles in either case. Each bottle is given its own cozy PVC home. This ensures that the bottles don't smack against each other and provides ironclad protection against breakage. You can adapt the rack a bit to suit whatever wine you'll be storing. The four tubes here are meant to accommodate both standard

750-milliliter bottles and wider bottles of sparkling wine or champagne. If you want to dedicate the rack to just white or red, you can use three-and-a-half-inch PVC pipe instead of the four-inch width specified here. That will allow you to add a pipe and increase the capacity of the rack by one bottle. You may also want to customize your rack if you want to secure the lid on top during transport (the design below allows for the necks of the bottles to stick up above the top of the bucket, making it easy to grab them). If you do want to use a lid, you'll need to remove one layer of the bottom insulation board, so that the bottles sit lower and the top fits neatly over the rim of the bucket.

No matter what you're storing, though, if you're keeping the wine in the same place for longer than a week, store the rack on its side to ensure the wine corks don't dry out. Also keep the rack out of direct sunlight whenever possible. The sun's rays can degrade the wine. That's a good reason to paint the bucket—which is also your chance to decorate it so that it pleases the eye as much as the palette.

PROJECT OPTIONS

One of the big advantages to this particular wine rack is portability. It allows you to enjoy a bit of sophistication on a long camping trip or to bring the beverages for a large group sitting on the grass for a concert in the park. Of course, wine lovers love more than just red wine.

• If you want to bring white or sparkling wine, you can chill the bucket by adding ice in the outside cavities. In conjunction with the foam, this will chill the bottles sufficiently to be tasty and refreshing.

• You can adapt the wine rack as a permanent location for your wine collection by crafting several of the wine racks and then placing them side by side using the cradle described on page 28. You can stack extra five-gallon-bucket wine racks on top, laying them between the bottom buckets.

WHAT YOU'LL NEED

Time: **30 minutes** | Difficulty: **Easy** | Expense: **$**

TOOLS:

Table saw or hacksaw

Measuring tape

Sharpie

Utility knife

Mallet (optional, for tapping pipes down into the bucket)

MATERIALS:

4" PVC pipe (42" section)

5-gal. bucket

Sheet of 2" foam insulation board (2×2" or 4×8")

Expanding foam sealant

HOW YOU MAKE IT

1. Cut the PVC pipe into 4 pieces 10½" long. Use the bucket bottom as a template to outline 2 equal circles out of a sheet of foam insulation board. Cut the circles out with the utility knife, cutting slightly inside the marked line. Stack the board circles in the bottom of the bucket.

2. Push the pipe sections down into the bucket in a cloverleaf pattern. Because the bucket narrows toward the bottom, you may need to tap the pipe sections down. The top edges of the pipes should be even with the top edge of the bucket, and the bottoms of the pipe should be in contact with the top foam circle.

3. Spray the expanding foam sealant down at the base of the pipes, in the spaces between the pipes. Spray only a little at a time, allowing it to fully expand before spraying more and holding the pipes down in the bucket as necessary. Spray in enough sealant to fill about 2" of each space. Repeat with each cavity. The pressure created by the sealant between the pipes and the bucket surface will hold the pipes in place.

Alternative: If the pipes squeeze up before you can spray the cavities with the expanding foam, you'll need to secure them in place. To do this, secure the lid on the bucket over the pipes and drill 1" holes over the cavities between the pipes. Use the holes to spray the expanding foam sealant into the cavities between the pipes, working slowly and carefully.

1.

3.

HANDY HOME ADDITIONS

Variation: Portable Beer Cooler and Storage

It's easy to adapt this project to beer bottles if your tastes tend more toward suds than wine. Because the diameter of a beer bottle is smaller than that of a wine bottle, you can fit more openings in this carrier than you would in a wine rack. Also, because you won't be storing the beer on its side, the cooler is crafted so that the tops of the bottles sit below the top edge of the bucket. This means the lid can be secured on top for transport or just to keep the bottles cold.

On that note, there are seven openings—to accommodate a six-pack and space in the center for ice. You can also use the cooler for beer cans, two per opening. However, you may want to use a couple long pieces of fabric for each pipe—pop a can down on the fabric and it will be easier to pull out when the thirst hits.

WHAT YOU'LL NEED

Time: **30 minutes** | Difficulty: **Easy** | Expense: **$**

TOOLS:

Sharpie

Table saw or hacksaw

Measuring tape

MATERIALS:

Sheet of 2" foam insulation board (4×8")

5-gal. bucket

3" PVC pipe (70" section)

Expanding foam sealant

HOW YOU MAKE IT

1. Cut the foam insulation board as in the wine rack project, and place the disks in the bottom of the bucket.

2. Cut the PVC pipe into 7 sections, each 10" long.

3. Arrange the pipes in the bucket with one in the center and six spaced equally around the outside (they should sit about ½" below the top edge of the bucket). As with the wine rack, spray about 2" of expanding foam at the base of the pipes, in the spaces between them.

Camping and Composting Toilet

Camping can be incredibly fun, but the least fun part of any outdoor adventure is going to the bathroom. That's because there isn't any bathroom. Unless you happen to be car camping in a campground equipped with facilities, or driving your own well-appointed RV, the bathroom is going to be wherever you can find a private place. But let's face it: squatting behind a tree, even for the experienced outdoorsman, is anything but pleasant. What you need is a toilet you can bring with you. Ideally, that fixture should be comfortable, convenient, and have no impact on the environment, and it must not add unpleasant smells to the campsite. That's an awfully tall order to fill.

The answer, of course, lies inside a five-gallon bucket.

The toilet described in the steps that follow is light and portable, simple to construct or break down, comfortable and easy to use—for children as well as adults—and is environmentally friendly to boot. Because you can't flush it, the secret lies in the absorbing power of recyclable elements. A container of sawdust takes care of most of the smells and ensures the toilet doesn't draw insects or create an unpleasant odor. If you don't happen to have access to sawdust, you can just as easily use a large amount of used coffee grounds.

In any case, by using a biodegradable, compostable trash bag to line the toilet, you ensure that the bucket is kept clean and that the waste can be placed in a hole and covered. It's the guilt-free water closet because the waste, including the bag, will naturally break down over time.

Always check local regulations about disposal of waste—many organized campsites don't allow it. If you dispose of the waste in a hole on your property, make sure it's not near where you grow any edibles, including fruit trees. The compostable bag—known in the roughing-it community as night fertilizer—is just as good as animal manure for enriching the soil of ornamental plantings.

HANDY HOME ADDITIONS

TOOLS:

Utility knife

Heavy-duty wire cutters

Clamp

MATERIALS:

½" polyethylene pipe wrap insulation

Large compostable garbage bags (10 gal. or larger)

Sawdust (enough to fill an empty plastic 2-lb. coffee container)

5-gal. bucket with lid

Roll of toilet paper

HOW YOU MAKE IT

1. Use the utility knife to cut the pipe wrap insulation to a length roughly equal to the circumference of the bucket's rim. Fit the insulation over the rim and adjust the length as necessary.

2. Slide the plastic handle to one side and cut the wire as close to the center as possible.

3. To set up the toilet, pull the cut part of the handle apart, slide the toilet paper down onto the handle's plastic tube, and slide the tube over the cut. Line the toilet with a compostable garbage bag. When using the toilet, clamp the pipe-wrap "seat" onto the rim, which will hold the bag in place. After use, sprinkle a generous amount of sawdust (or even dirt from the campsite) into the toilet, remove the pipe wrap insulation, and secure the lid on the bucket. When full, remove the bag and bury or dispose of it according to local regulations.

PROJECT OPTIONS

Sitting on pipe wrap insulation is a whole lot more comfortable than sitting on the bare rim of a five-gallon bucket, but it is admittedly not as comfortable as sitting on a standard toilet seat. If your campout gang is more into comfort, you can adapt a standard "round" toilet seat for use on this camping toilet. Attach two three-inch sections of the pipe insulation to opposite sides of the bucket rim. Center the toilet seat over the rim and note where it contacts the pipe insulation. Coat the top of the pipe insulation pieces and the contact areas on the bottom of the toilet seat with adhesive meant specifically for plastics—such as J-B Weld's PlasticWeld (see **Resources**, page 144). Let the seat sit on top of the insulation until the adhesive dries. Now you have a comfortable seat that can simply be attached and detached when using the toilet.

Dryer Lint Trap

Dryer lint is a small thing that can have big impact. Improperly exhausted dryer lint accounts for tens of thousands of house fires each year, resulting in millions of dollars' worth of damage. And that's just the really scary facts. Even if you fastidiously clean your lint catcher in the dryer, some lint will make it into the venting ductwork. Improperly routed ducting, incorrectly sized vents, and blocked vents can also be the source of problems. Those issues can raise humidity levels around the vent itself, causing moisture to accumulate and mold to grow in the adjacent wall cavity. Lastly, lint that even partially blocks an exhaust hose or vent can drastically decrease the efficiency and lifespan of your dryer.

An efficient external lint trap can be a safety device and insurance against wear and tear on the unit itself. Most dryers are vented to the outside of the home through a screened hole in the wall. This means the dryer has to be located against or near an outer wall. A self-contained lint trap allows you to put the dryer wherever it is most convenient for you, including a basement or utility closet. A water lint trap, such as the one described here, also traps heat from the dryer. You can use this to your advantage in cold weather, to lower your heating bill.

This trap is actually very simple. The dryer vent hose is routed to a five-gallon bucket filled about one-quarter full of water. The water catches and holds most of the lint. A screened section of the lid allows for efficient airflow but prevents any residual lint from flying out of the bucket.

You'll want to make a quick check of local codes—at the zoning and fire departments—to ensure that a self-contained lint trap is permissible in your area. It's also not a wise idea to use an external lint trap so that you can add a washer and dryer to an interior space in a rental, like a closet or alcove. Lastly, under no circumstances should you use this trap with a gas-powered dryer, because it could lead to the buildup of dangerous carbon monoxide.

Time: **25 minutes** | Difficulty: **Easy** | Expense: **$$**

WHAT YOU'LL NEED

TOOLS:

Sharpie

Cordless drill and bits

Jigsaw

Tin snips or heavy-duty scissors

Screwdriver

MATERIALS:

5-gal. bucket with lid

Fiberglass replacement screening

Silicone sealant

4" PVC male transition fitting, Schedule 40

5" hose clamp

HOW YOU MAKE IT

1.

1. With the lid fastened on top of the bucket, use the end of the PVC transition fitting as a template to mark a vent hole in the lid of the bucket with the Sharpie. Drill a starter hole and then use the jigsaw to cut out the hole.

2. Remove the lid and use the jigsaw to cut a half-moon section opposite the hole. This section should be as large as possible. Use the cutout in the lid as a template to mark the half-moon cut on the roll of screening. Use the tin snips or scissors to cut out the half-moon of screening ½" larger all around than the line marked.

2.

3. Lay a generous bead of silicone sealant around the underside of the half-moon cut in the lid, and lay the screen in position. Allow the sealant to fully dry.

4. Lay a bead of silicone around the rim of the vent hole on top of the lid. Set the PVC fitting into place, pushing it through the hole from the top, and laying the flange in the silicone. Allow the silicone to dry.

5. Add 2 gal. of water to the bucket and secure the lid on top. Connect the flexible duct hose from the dryer over the male adapter, using the hose clamp. Dry a load of wet clothes to ensure that air flows freely through the trap and that it is capturing most of the lint.

4.

QUICK 3
Three Quick and Useful Bucket Options

Whip together a few handy home additions to repurpose any five-gallon buckets lying around. These are all simple solutions to clutter, requiring little time, effort, or expense.

1.
Plastic Bag Dispenser

If you're like most people, you have a loose collection of plastic shopping bags. Don't throw them out or let them become a sprawling mess. Cut a hole in the side of a five-gallon bucket, shove them all inside, and you have a handy dispenser.

2.
Hose Handler

Keeping garden hoses in order can be a challenge. To create a hose handler, fasten a bucket onto an inside or outside wall with the bottom of the bucket flat on the wall stud using just a few screws. It provides space inside for sprinklers, gloves, or other gardening implements. Attach it to a garage wall or fence near your outdoor spigot to always keep the garden hose in order. Or use an alternate method—fill it with rocks and wrap the hose around it!

3.
Bucket Spacer

Making the most of this book means collecting five-gallon buckets whenever you can. Keep your buckets organized in stacks with simple spacers made of empty liter bottles with the lids screwed on. As long as the bottles are sealed, the air stays inside and they will keep buckets from sticking together due to suction.

Simple Storage Stool

A portable stool is one of the handiest types of seating. Although this was conceived as a comfortable padded place to take a load off in the workroom or campsite, the stool can also be used at a home office desk, around an outdoor picnic table in place of bench seating, and just about anywhere else you need a place to sit but where a chair would be overkill.

The bonus with this particular stool is a goodly amount of storage space right underneath the seat. Who can say they ever have too much usable, accessible storage? It's best to exploit this roomy space as long-term storage. Frequently snapping the lid on and off the stool could lead to wearing out the fabric that is wrapped around the lid for the seat. It's also a great way to secure food at a campground, keeping it safe from hungry wild animals looking for a free snack.

As far as what fabric you use for the seat, the sky is the limit. Of course, the more durable that fabric, the better. (Once you put this stool to work, you'll be using it all the time.) Linen and denim are ideal choices for the stool seat—you could even recycle an old, worn-out pair of jeans or a denim skirt. If, however, you intend for the stool to see less use (if you're pairing it with a dressing room vanity, for instance), you can opt for a nicer-looking, more delicate fabric. Whatever fabric you decide on, it's a good idea to ensure that it is cleanable (especially if it's going to be used in a workshop or kid's room). If it isn't naturally water and stain resistant, you might want to use a protectant spray to keep the seat looking nice for as long as possible.

Lastly, you can make the stool even more attractive by painting it a solid color or in a pattern of some sort. Choose a hue and design that matches whatever room the stool will be used in. For instructions on the best way to paint the stool, see **Painting Your Bucket**, page 56.

PROJECT OPTIONS

You can make this project even simpler (if a little more expensive) by purchasing one of several aftermarket lids made with integral seats (see **Resources**, page 144). Some are merely a padded top on a lid that replaces the bucket's lid, while others are more customized, with leather or vinyl top surfaces, plush filling, and even special shapes to accommodate the sitter.

TOOLS:

Sharpie

Scissors

Fid or knife

Staple gun and staples

MATERIALS:

5-gal. bucket with lid

1" foam padding

Fabric (approximately 20" square)

Fabric adhesive

Cotton batting (sheet or loose fill)

HOW YOU MAKE IT

1. Remove the lid and make sure it is clean. Use the lid as a template to mark the foam for the seat with a Sharpie. Cut out a foam circle the exact size and shape of the lid. Use fabric adhesive to glue it down to the top of the lid (or use adhesive appropriate for the foam you're using).

2. Lay the fabric square out on a clean work surface. Mound the cotton batting (or a circle cut from sheet batting) in the center of the fabric. Dot fabric adhesive on the foam covering the lid, and press it down (upside down) centered on the mound of batting.

3. Pull two opposite sides of the fabric over the edges of the lid, so that the fabric is taut on the top of the seat. (Use a helper for this step.) Holding the fabric tight, use the fid or the back of the knife to press the fabric down into the channel of the lid. Then staple the excess fabric to the underside of the lid.
   Alternative: If securing the fabric in the lid channel seems a bit difficult, you can staple the fabric to the outside rim of the lid and then double it up into a hem, and glue the hem over the staples.

4. Paint the bucket if desired and fasten the lid in place. If the bucket wobbles, you can add weights inside to stabilize it. Remove the handle for a nicer look, or leave it on if you'll be moving the stool around quite a bit.

2.

3.

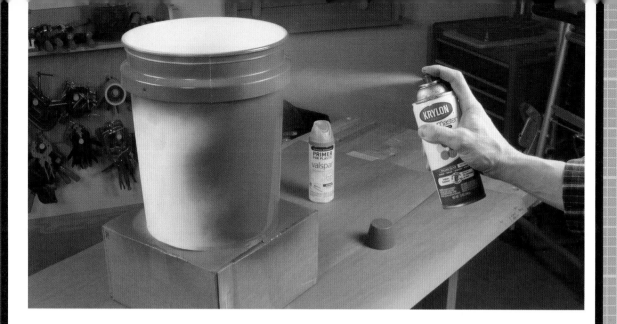

PAINTING YOUR BUCKET

It's the rare five-gallon bucket project that can't be spruced up with a coat of colorful paint. Start by selecting the paint. The best way to ensure a lasting paint job that looks good now, and later, is to use a paint labeled for plastics, such as the Fusion for Plastic spray paint by Krylon (see **Resources**, page 144). In most cases, the paint will be a spray paint. You'll find the widest selection of paint colors at large home centers and paint stores. Follow these steps to ensure success.

1   Stage properly. Set up your work area with proper ventilation, and lay plastic over anything—including the floor—that you don't want painted. Make sure you have the proper safety gear. Use a respirator rated for spray paints, safety glasses, and work clothes that cover most of your skin and your hair. Lastly, be sure you have enough paint on hand and 100-grit sandpaper. It also helps to have a solvent meant for use with aerosol paints, to clean up any mistakes.

2   Prep the bucket. Clean the bucket thoroughly, so that there is no grease or dirt on it. Sand as necessary, according to the manufacturer's instructions for the paint you're using. Mask off any areas on the bucket you don't want painted.

3   Paint carefully. Prime the bucket as needed. Most spray paints meant for plastics don't need a primer coat to adhere. However, it's never a bad idea to prime with a paint meant to be used with the top-coat paint. Follow the instructions on the can. Always keep the spray tip in motion to avoid drips, and lay down a very thin coat for the same reason. You can always add a second coat.

4   Add patterns as desired. Decorate with designs by spraying a top coat of one color, masking the design (or sticking stencils down for figural designs), and spraying another top coat in a different color. For the best appearance, sand very lightly (with 180-grit sandpaper) between top coats and use a sealer to protect the paint and add sheen.

Rod-and-Tackle Bucket

Keeping your fishing rods, and all that goes with them, safe and organized can be a bit of a challenge. Stand them in a corner and they're likely to fall down. Lay them down anywhere and there's a good chance that something heavy will be piled on top of them, breaking some prized outdoor tools. The fishing lines make things even more complicated because they can come loose and tangle—especially where you're storing multiple rods together. Protecting the rods and accessories such as reels from damage is no small matter—breaking any of those things can set you back quite a bit, regardless of whether you repair or replace it.

There is an easy way to protect your rods and other fishing gear, and keep everything neat and tidy both at home and at the lake. The handy rod-and-tackle bucket described here offers plenty of storage space—room enough for six rods—including space for a small tackle box, extra reel, rolled-up waders, or other gear (or even ice and fish if you run out of room elsewhere).

The great thing about this storage bucket is that it will stand up to a beating. Secure it in the back of a pickup and your rods will stay in one piece, even if something heavy slides into the bucket. The handle means transporting your gear to even remote fishing holes is a breeze. That handle also means you can hang the bucket from a tree branch or post at a campsite to keep the rods out of the way of foot traffic, and you can do the same at home if you want to store the rods from a rafter in the garage or shed.

One of the best things about this bucket is how easy it is to clean. Spend a long weekend fishing and things tend to a get a little grimy and smelly. But blast out the bucket with a hose and a little dish soap, and you'll be good to go on your next fishing trip.

TOOLS:

Circular saw or hacksaw

Measuring Tape

Sharpie

Hot glue gun and glue

MATERIALS:

2" PVC pipe (84" section)

5-gal. bucket without lid

Cordless drill and bits

(12) 2" flathead screws

HOW YOU MAKE IT

1. Cut 6 sections of the PVC pipe, each 14" long. Measure and mark 6 points equidistant around the inside of the bucket. Make sure the pipe sections and the inside of the bucket are clean of any dirt, grease or other coating.

2. Coat a stripe from top to bottom along each of the pipe sections with hot glue. Press the pipes into place along the inside of the bucket, and hold each for a minute or so to ensure a good bond. You can also jam a bag filled with sand down into the center of the bucket, if necessary, to hold the pipes in place while they dry.

3. Once the glue has dried, drill pilot holes through the outside of the bucket into each pipe, approximately 5" down from the top and 3" up from the bottom. Secure the pipes with 2" flathead screws driven into the pilot holes.

PROJECT OPTIONS

You can make your rod-and-tackle bucket even more useful by spending a few dollars on a rod holder that attaches right to the bucket for hands-free fishing. These aftermarket add-ons are inexpensive and can make for a peaceful afternoon fishing without fuss (see **Resources**, page 144).

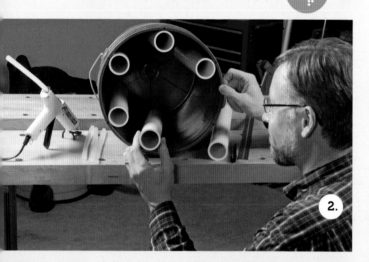

2.

3.

Child's Bucket Seat

Any toddler would be delighted to have his or her very own bucket seat. It will become a wonderful place to sit at a low table coloring or playing with blocks, and it can even be moved to the living room so that youngsters can have their very own place of honor to sit and watch their favorite TV programs or videos. But given the durability and ruggedness of a five-gallon bucket, you can also turn this into a nice place for your little one to sit outside, enjoy a little sun, and get the daily dose of vitamin D as he or she reads a picture book.

Although the construction of this tiny seat is not difficult, you will have to pay attention to the details. It's especially important to completely sand all the cut edges smooth. You don't want a sharp plastic edge cutting your child's leg or back. There are also a number of ways to add to the comfort of the seat. You can easily increase the seat's padding to make it even more comfortable and, instead of the rim padding outlined in this project, you can make a fully padded back by gluing down padding across the entire surface and then wrapping it in fabric.

You can also customize the seat to suit the sitter. Pick out fabric with your child's favorite superhero or cartoon figure, or just use a brightly colored pattern. You can also paint the base to match (see **Painting Your Bucket**, page 56), complement or contrast the fabric, and stencil it with designs such as stars or even the child's name. You can even adapt the project to make the seat a bit more useful by making it removable, which will open up the space underneath (see **Smart Seat Mod**, page 61).

Time: **1 hour** | Difficulty: **Moderate** | Expense: **$$**

WHAT YOU'LL NEED

TOOLS:

Sharpie

Straightedge

Measuring tape

Cordless drill and bits

Jigsaw

Carpenter's compass

Staple gun and staples

Hot glue gun and glue

MATERIALS:

5-gal. bucket

80-grit sandpaper

100-grit sandpaper

½" pipe wrap insulation

Fabric (16" square)

Fabric adhesive (liquid or spray)

½" plywood scrap (at least 14" square)

2" foam sheet (at least 14" square)

Loose cotton batting

2" flathead wood screws

2.

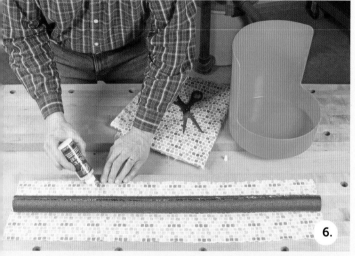

6.

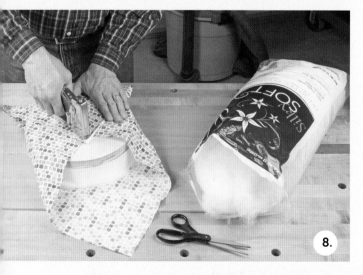

8.

HOW YOU MAKE IT

1. Use the Sharpie to mark a line under the lowest ridge on the bucket, all the way around to behind the handles on both sides. Use a straightedge to draw a line from this top line, down to about 3" above the bottom of the bucket, on both sides.

2. Measure and mark points around the front of the bucket, 3" up from the bottom. Mark a point every 1 to 2" around to the opposite vertical line. Use a flexible straightedge to draw a line through these marks connecting the two vertical lines. Draw a gentle curve at the lower corners where the bottom horizontal line meets each vertical line.

3. Drill an access hole at one corner where the vertical line meats the top horizontal line. Use this hole to start the jigsaw cut. Follow the line as closely as possible with the jigsaw, across the back, down one side, across the front, and then up the other side.

4. Sand the cut edges of the seat, starting with 80-grit sandpaper and progressing to 100-grit. Make sure all the edges are smooth and free of any snags or sharp points.

5. Cut one piece of pipe wrap insulation long enough to run up one side of the back, around the top of the back, and back down the other side of the back. Cut a long thin piece of fabric to wrap around the insulation. The fabric should be wide enough so that the ends can be folded up into the wedge of the pipe insulation slit.

6. Lay the pipe along the center of the fabric. Apply a line of fabric adhesive along the fabric edges at each side. Wrap the fabric up on each side, folding the ends into the slit, and pressing tightly along the fabric adhesive. Set aside to let dry.

7. Measure the diameter of the seat, from the front cut edge to the back wall. Transfer this measurement to the plywood scrap, and draw a circle using the compass. Cut out the plywood circle with a jigsaw, using the method described on page 23.

8. Use the plywood circle as a template to cut a corresponding foam circle. Cut a large fabric square for the seat, about 4" larger than the diameter of the foam. Glue the foam to the plywood with fabric adhesive. Mound cotton batting on top of the foam and then lay the fabric over the top. Collect the fabric at the edges, flip the seat, and, holding opposite sides, staple the edges of the fabric with the fabric held snug (it is much easier to do this with a helper).

9. Snug the seat down level inside the bucket. Drill 6 pilot holes equidistant around the perimeter of the bucket, into the edge of the plywood (A). Drive wood screws through the holes to secure the seat. Lay a bead of hot glue along the inside edges of the fabric-wrapped pipe insulation and press it into place (B). Let the glue dry before letting your child use the chair.

9A.

9B.

SMART SEAT MOD

It's easy to make this handy chair even more useful by making the seat removable so that you can use the space underneath for storage. Instead of screwing the seat in place, measure and cut three two-by-two-inch pieces long enough to run from the bottom to one-half inch below the edge of the chair. Place the supports equidistant around the inside of the bucket, and drill pilot holes through the outside and into the wood. Screw the bucket to the wood from the outside, using two-inch flathead wood screws. Sew or glue a small fabric handle to the back edge of the seat, and push it down into place over the supports. Now your child can hide a favorite toy or other precious items inside a comfy chair.

Toy Storage Center

Everyone loves their children, but toy clutter is another matter altogether. It is amazing how many toys children can accumulate, and how many types—and shapes and sizes—there are. Those toys seem to take on a life of their own when it comes to making a mess. Scattered toy cars, blocks, and marbles are probably not the look you had in mind for your home when you picked out the paint and furniture. And it doesn't help that those harmless little playthings can be a danger underfoot. Sadly, getting the kids to pick up after themselves is perhaps the biggest part of the challenge of keeping your living room or play room uncluttered when it's not playtime.

Fortunately, kids can be trained (or coerced, rewarded, or punished) to pick up the mess. Getting them in that habit will be way easier if you have a handy—not to mention fun—toy storage center. The one described in this project provides plenty of room for toys of all kinds—and shapes. Children will find it simple and even a bit fun to literally toss their toys into the storage buckets.

Your room design being of paramount importance, it helps that this particular storage unit is beyond easy to customize. You can paint it and decorate it to blend right in, or go fun and fabulous and make it a centerpiece all its own. Painting or stenciling the framework is as easy as painting the bucket (see **Painting Your Bucket**, page 56). As a bonus, the project will go together with a bare minimum of tools and modifications. The supporting framework has been designed to use off-the-shelf plumbing parts. And the benefits don't stop there.

Because the whole structure is crafted of high-density plastic, spills, dirt, and even minor abrasions are a breeze to clean up.

The unit is also scalable. Use the basic principles and parts to expand this storage center to a six-, eight- and even ten-bucket home for toys as needed. If you locate it in a kid's bedroom, designate one bucket for laundry by lining it with a mesh laundry bag and putting a toy basketball hoop over the front of the bucket. You'll kill laundry clutter along with the toy mess!

Time: **1 hour** | Difficulty: **Easy** | Expense: **$$**

WHAT YOU'LL NEED

TOOLS:
Cordless drill and bits

MATERIALS:
(2) ½×12" slip-fit nipples (Schedule 80)

(10) ½×10" slip-fit nipples (Schedule 80)

(4) ½" slip-fit elbows (Schedule 40 or 80)

(5) ½" slip-fit tees (Schedule 40 or 80)

(3) ½" slip-fit snap (saddle) tees

(2) ½" couplings (Schedule 80)

PVC cement

(4) 5-gal. buckets without lids

Duct tape

#4×½" flathead wood screws

(2) ½" slip-fit caps (Schedule 80, optional)

2.

4.

5.

HOW YOU MAKE IT

1. Lay out the PVC frame pieces on the floor, in the position they go together. Working from the bottom, assemble the bottom and first row of vertical supports, connecting the pieces with PVC cement.

2. Make a 4-way by snapping a snap tee on top of the middle tee fitting. Slip the top middle vertical support into the snap tee before connecting it.

3. Lay the completed frame on the floor to check alignment and that the frame lies flat. Adjust as necessary.

4. Place the buckets upside down, as they'll be in the frame. Slide the frame, front-side down, over the buckets. Press the frame down and duct tape the bottom cross braces and side braces to each bucket.

5. Flip the unit so that the buckets are right-side up. Drill pilot holes, and drive two ½" screws into the bottom cross brace and two side braces for each bucket.

6. Cement a cap onto one end of each 12" nipple. Slip a snap tee on the opposite end. Position one of these legs on each outside leg. Stand the frame up. To make it stand more upright, move the snap tees down (add a coupling and 4", 6", 8", or 10" nipples, as preferred).

6.

TOY STORAGE STRATEGIES

A toy storage center like this one is only one part of a clutter-free room. To make the most of the storage, you need to motivate children to pick up after themselves. It helps if the storage center is painted in bright, fun colors. Engage your children to pick out the colors, and let them help you paint the frame and buckets.

• Stencil the names of the types of toys to be stored in each bucket. This is a great way for kids of reading age to clearly know where to put what. Emphasize to the children that when things are organized in this way, it's much easier to find their favorite toys.

• For children who don't read, take a picture of each type of toy, print it out, and tape the pictures over the bucket for that type of toy.

• Make a point system. Each time mom or dad come into the room and the buckets are full of toys with none left on the floor, the kids get a certain number of points. Anytime toys are left out, points are deducted. When the kids accrue enough points, they get a simple treat.

• Drill air holes in any buckets that hold soft, absorbent toys. This will prevent mold and smells that might keep children from using the bucket for storage.

• Make one bucket just for art supplies. These supplies tend to be messy and hard to find when it comes time to do an art project for school or fun. A single bucket dedicated to arts and crafts confines the mess and makes it easy for children to find what they need.

• Dedicate one bucket to toys for donation. Periodically have your children go through their toys and pull out those they don't play with anymore. These "past their prime" toys go in the donate bucket to be given to charity. That clears space for new birthday and Christmas presents, ensuring there aren't more toys than the storage center can accommodate.

Handy Workbench

There are few things more useful to the home craftsman (or anyone else, for that matter) than a sturdy, portable work surface. The right workbench is a place to create, craft, fabricate, design, and dream. It should be rugged and durable so that heavy or rough materials won't damage it—the occasional coffee-cup spill or accidental drill-through should not be a big deal. And it should be the right size to fit in a garage or workshop without taking up too much room (but still providing enough space to work comfortably). It should almost be like another tool and quickly become indispensable. That's a lot to ask of one work structure.

The five-gallon bucket delivers. This project, and the specialized variations that follow, provide an easy-setup, easy-breakdown work surface that can be the center of a busy workspace. It serves just as well inside as it does out. Not only will it be the perfect place to do some drilling or sawing, it also makes a wonderful potting table in the yard or garden. It's even great as a staging area for indoor home improvement projects, such as prepping and painting an entire room. (Of course, don't forget, it could serve as the launch pad for many a five-gallon bucket project!)

Portability is a key benefit to this table. You can take it apart in minutes and transport it with ease. That means it can double as a buffet table at a large family cookout or head to the nearest school fundraiser to serve as the base for a display of baked goods. The table can be put together almost entirely from recycled building goods, including empty drywall compound buckets and a beat-up sheet of plywood. Even if you have to buy the components, the low cost is yet another attractive feature of this hardworking surface.

The variations on pages 68 and 69 are twists on the main theme outlined in the steps that follow. They allow you to customize this work surface, making it serve your needs no matter what it is you want to do. In any case, though, the surface will be easy to assemble and cheap—much less expensive than buying a workbench or even two high-quality sawhorses.

PROJECT OPTIONS

If you want to make the table legs more stable, add weights—such as filled water jugs, bags of sand, or other heavy objects—to the bottom bucket of each leg. To make the table capable of supporting more weight, turn the bottom bucket of each leg upside down and screw the bottoms of both buckets in each leg together.

TOOLS:

Circular saw or table saw

Measuring tape

Sharpie

Vise or clamp

Jigsaw or hacksaw

MATERIALS:

(8) 5-gal. buckets with lids

Sheet of exterior-grade ½" plywood

HOW YOU MAKE IT

1. Use the circular saw or table saw to rip and cut the plywood down to 20½" wide and 72" long (or use the full 8' length if you prefer a much larger workbench). Measure in 2½" from the edge of one bucket lid and mark with the Sharpie. Use the straightedge to mark a straight line dividing the lid in unequal sections, at the mark.

2. Hold the lid in the vise or clamp it to a work surface so that the cut line is not obstructed. Use a jigsaw or hacksaw to cut the lid along the cut line. Repeat the process with three of the remaining lids.

3. To assemble the table, stack one of the buckets with a cut lid on top of a bucket with an uncut lid. Stack two buckets in the same way next to the existing stack. Rotate the top buckets so that the cut portions of the lids align on the outside edges of the buckets. Repeat with the other buckets to create two opposite legs, centered about 4' from the original legs. Set the tabletop in place and move the legs as necessary for better support.

1.

2.

Sawhorses are super-handy workshop helpers, adaptable to just about any time you need a work surface right now, right here. This design is better than most because the "legs" are added storage, helping you lug whatever you need to the job site. Those legs also add a work surface at the end of each crossbeam, making this sawhorse even more useful. Make as many multiples of this design as you might need for a given worksite—they're easy, quick, and inexpensive.

Time: **20 minutes** | Difficulty: **Easy** | Expense: **$**

WHAT YOU'LL NEED

TOOLS:
Sharpie
Cordless drill and bits
Jigsaw
Measuring tape

MATERIALS:
2×4 (ideally 6' long, but size as desired)
(4) 5-gal. buckets with lids

HOW YOU MAKE IT

1. Turn one of the buckets upside down on a work surface. Lay the 2×4 on edge across the center of the bucket bottom, and use it as a template to mark cut lines with the Sharpie. Remove the 2×4 and extend the cut lines 3½" down either side of the bucket. Connect the ends of the cuts on each side with a short horizontal cut line.

2. Drill an access hole with the cordless drill, then cut along the cut lines with a jigsaw to remove the section of the bucket bottom in which the end of the 2×4 crossbeam will rest. Repeat with a second bucket.

3. Assemble the sawhorse on a flat surface by stacking the buckets in legs upside down, with the cut buckets on top. The board gaps in the buckets must align. Lay the 2×4 crossbeam in the legs and move the legs as necessary to fully support the 2×4.

Sometimes, all you need is a small work table to hold a piece of wood or a bucket of paint for a moment. In those situations, there's nothing like a "workmate." Set this up in the blink of an eye, and it takes up about as little space as possible, while still providing a stable, modest work surface.

Time: **20 minutes** | Difficulty: **Easy** | Expense: **$**

WHAT YOU'LL NEED

TOOLS:

Measuring tape

Sharpie

Cordless drill and bits

MATERIALS:

2×2' plywood scrap (at least ¾" thick)

(2) 5-gal. buckets with lids

(3) #4×½" flathead screws

HOW YOU MAKE IT

1. Determine the bottom of the plywood scrap, and measure and mark two perpendicular centerlines. Lay the plywood bottom-side up on a clean, flat work surface.

2. Lay one of the bucket lids top-side down on the plywood, and center it using the centerlines. Drill 3 pilot holes equidistant around the center of the lid, in a triangle. Drill through the lid and into the plywood. Drive the ½" screws through the holes, securing the lid to the plywood.

3. To create your quick-action workmate, snap the lid with attached work surface on one bucket and stack it on the lid of the other bucket.

BARNYARD HELPERS

2

IT DOESN'T MATTER IF YOU HAVE A HOMESTEAD FARM, A BARNYARD, A BACKYARD, OR A LARGE GARDEN; IF YOU'RE RAISING ANIMALS TO USE AS FOOD OR PRODUCE FOODSTUFFS, THE INNOVATIONS IN THE PAGES THAT FOLLOW ARE SURE TO MAKE YOUR LIFE A LITTLE EASIER AND THE ANIMALS A LITTLE MORE COMFORTABLE. THE FOCUS OF EACH INNOVATION IN THIS CHAPTER IS TO PROMOTE ANIMAL HEALTH, SAFETY, AND WELL-BEING, FIRST AND FOREMOST. THAT, IN TURN, ENSURES THAT THE FOOD YOU GET FROM YOUR ANIMALS IS THE HIGHEST QUALITY.

Raising farm animals can be a labor-intensive pursuit. That's why all of these projects have been designed to cut down on the animal-care chores, without shorting the animals themselves. All of these are fairly easy to execute and will take up a bare minimum of your time or energy. The payback in terms of chore time saved should make any effort more than worthwhile. The projects included deal with animals small and large, because even if it's beyond your capacity to raise pigs, a hive of bees is doable for just about any garden.

It's important to mention that animals are living organisms, so just as you wouldn't recycle an industrial cleanser five-gallon bucket to make something your kids will interact with or that you'll make food or drink in, you shouldn't use dirty or contaminated buckets for any of the projects in this section. New general-purpose buckets are fine, but recycled buckets should be food grade and should have previously held foodstuffs.

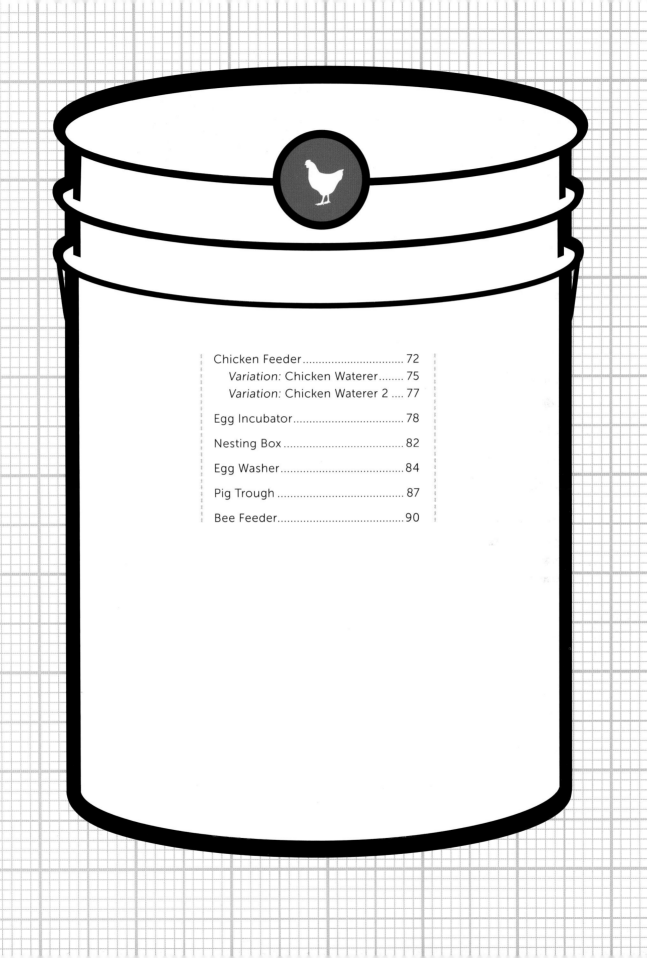

BARNYARD HELPERS

Chicken Feeder

Feeding chickens everyday isn't the worst chore ever, but it doesn't really need to be a chore at all. You can save some time and effort by constructing a fill-it-and-forget-it chicken feeder. You'll only need to fill this handy clucker accessory about once every seven to ten days for a small flock of ten chickens. The trick to the feeder is to let gravity do the work. And because the bucket used in this feeder is positioned right-side up, you can use the handle to hang the unit up off the ground (the preferred strategy for a number of reasons—see **Safety First** on page 77). You'll need a clean bucket that has an opening in the lid equipped with a screw-on top, like the pour spout in a five-gallon bucket of paint. That way, you won't have to wrestle the bucket's lid off when it comes time to refill the feeder—just unscrew the small top and add chicken feed through a funnel. Make this at home following the steps below and you'll be saving a whole lot of money—store-bought versions of chicken feeders can run from $35 to $60!

You can save even more money by reusing found materials. Although we've used a plastic or rubber garbage can lid for both the feeder and the alternative projects that follow, a rubberized oil pan or even a roasting pan would work just as well. However, it's wise to stay away from materials that are prone to rust or corrosion. And keep in mind that chickens tend to climb on and jostle structures in their environment; the more rigid and durable any material you use for a feeder or watering basin, the better.

Lastly, whether you choose to hang the feeder or just sit it on the ground, we suggest you put it outside the coop to help keep the coop nice and clean. (Chickens are messy eaters!) The waterer, on the other hand, can go inside the coop so that the birds always have access to a fresh, clean water source.

Time: **30 minutes** | Difficulty: **Easy** | Expense: **$**

WHAT YOU'LL NEED

TOOLS:

Sharpie

Measuring tape

Cordless drill and bits

1½" hole saw

Crescent wrench

Standard screwdriver

MATERIALS:

Clean 5-gal. bucket with a threaded hole in the lid and a screw-on top

(3) ¼" stove bolts and nuts with washers

Metal, plastic, or rubber garbage can lid (no handle)

HOW YOU MAKE IT

1. Mark 4 holes 1" up from the bottom of the bucket, equidistant around the bucket's surface. Secure the bucket upside down, with a weight on the bottom (or clamp it in a large vise), and drill the holes using the 1½" hole saw.

2. Mark 3 holes on the top of the garbage can lid, in a triangle with each hole approximately 3" from the center. Drill ¼" holes at the marks. Center the lid over the upside-down bucket and mark corresponding holes on the bucket's bottom. Remove the lid and drill ¼" holes at the marks on the bottom of the bucket.

3. Attach the lid to the bottom of the bucket using 3 stove bolts, with the bolt heads on the lid side. Place a washer under both the head and the nut on each bolt. Tighten the nuts securely.

4. Turn the bucket right-side up, so that it's sitting in the upturned lid. Install a screw-top lid with separate frame, as shown here, or simply use a snap-on lid with a pour hole and cap. Unscrew the pour-hole screw top and fill the bucket with chicken feed. Let the chickens know it's dinnertime!

1.

2.

3.

4.

CHICKEN-FEEDING TIPS

Whether they're getting their feed from a feeder or just spread on the ground, chickens naturally scratch when they feed. This means that chickens may stand in a feeder on the ground, flicking food out while eating. If this becomes a problem, use a deeper feeder base, replacing the garbage can lid in this project with a plastic oil pan or similar container. Of course, you can also hang the feeder by its handle. Here are a few more tips to keep your birds well fed.

• If you're supplementing your chickens' feed with a calcium source, such as crushed oyster shells, put it in a separate feeder or dish. (Ironically, crushed eggshells are a wonderful calcium supplement for chicken feed!)

• When molting, chickens may need additional protein sources. Add mealworms to their food.

• Add one tablespoon of raw, unfiltered apple-cider vinegar to the chickens' water to eliminate potentially harmful bacteria.

Variation: Chicken Waterer

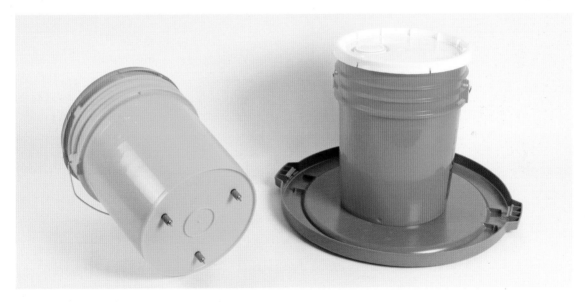

Chicken owners are increasingly turning away from traditional watering troughs to newer, more sanitary "nipple feeders." These feature simple nipple valves that the chicken pecks for a quick drink. Because the water is kept in a closed container and the nipple ensures a seal, many advocates consider this a good way to ensure the water source never becomes contaminated. Most chickens figure out the nipple system almost immediately. You'll be using push-in nipple feeders (these can be found at any large feed or supply store, or online—do not buy screw-in styles); follow the instructions supplied with your nipples. Use a chain and an S hook to keep the waterer off the ground so that chickens don't step in it and foul the water.

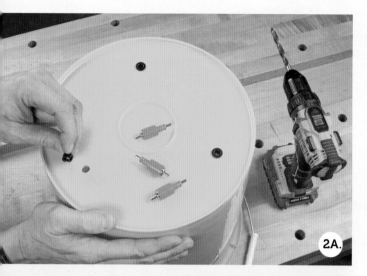

2A.

WHAT YOU'LL NEED

TOOLS:

Cordless drill and bits

MATERIALS:

5-gal. bucket with lid

(3) or (4) push-in style watering nipples with grommets

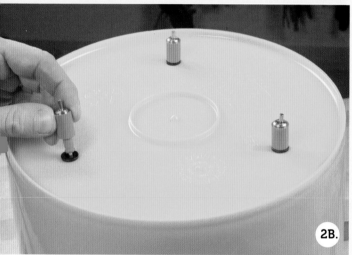

2B.

HOW YOU MAKE IT

1. Turn the bucket upside down and drill 3 to 4 holes distributed in a triangular or square pattern across the bottom of the bucket.

2. Wet one of the supplied grommets and push it into the hole (A). Then push a nipple in through the grommet until you feel it seat. Repeat with the remaining nipples (B) and then fill the bucket with water and check that the nipples are working correctly (a drop of water should come out when you depress the nipple).

SAFETY FIRST

To keep chickens healthy, make sure that they are not defecating in a feeder or waterer. Hanging is a sure solution to the problem, but regularly empty and thoroughly clean the bucket feeder and waterer to ensure against harmful contaminants such as mold. Keep waterers off the ground where there are chicks roaming freely. They can fall in and drown.

Variation: Chicken Waterer 2

If for any reason you can't find watering nipples or prefer not to use them, you can still keep your feathered egg machines hydrated. This innovation works on simple hydrodynamic principles, which require that the lid fit tightly, creating an airtight seal. (An airtight seal at the top is what prevents water from just flowing out of the holes and up over the rim of the base.) It's essential to use a bucket with a hole in the lid equipped with a screw-on top, so that you can easily refill the waterer. As with the feeder, the bucket handle allows this device to be hung.

WHAT YOU'LL NEED

Time: **30 minutes** | Difficulty: **Easy** | Expense: **$**

TOOLS:

Cordless drill and bits

½" bit

Crescent wrench

Standard screwdriver

MATERIALS:

Clean 5-gal. bucket with an opening and screw-on top in the lid

¼" stove bolt and nut

Plastic or rubber garbage can lid (no handle)

(2) rubber washers

Silicone sealant

HOW YOU MAKE IT

1. Secure the bucket upside down. Cut or drill a ½" hole in the side, right at the bottom edge. Center the garbage can lid over the bottom. Drill a ¼" hole through the center of the garbage can lid and into the bucket.

2. Secure the lid to the bucket with a stove bolt and nut. Use a rubber washer on both sides, under the bolt head and under the nut. Lay a small bead of silicone sealant underneath each rubber washer before tightening the nut, and bolt down as tight as possible without cracking the bucket.

3. Allow the silicone to dry. Hang the waterer in the most convenient area for the chickens (inside the coop if there is space). Use a hose to fill the bucket about ¾ full, through the hole in the lid. Quickly screw down the top on the fill hole. Check for leaks and fill with sealant as necessary.

1.

BARNYARD HELPERS

Egg Incubator

This project is a very basic incubator design. The point is to provide a good environment for the chicks-in-development while keeping costs low. The thrift translates to a minimum of effort that would otherwise be automated. That extra work gives you a chance to interact with the birds you're hatching and regularly check the health of the eggs. The capacity is modest, meant to hatch no more than five eggs at a time. That said, you can modify the design to match your preferences and expertise. It's easy to add an automatic egg turner, a thermostat with a switch to turn the lamp on and off based on temperature, and fans to circulate air and heat more evenly and efficiently.

The most important factor is a steady temperature. Chicken eggs incubate at about 99 to 102 degrees Fahrenheit. (If you want to incubate eggs of other species of birds or reptiles, investigate the ideal temperature for that particular species.) Humidity is important as well. The ideal humidity for the inside of a chicken egg incubator is about sixty percent. That means you'll need a water source inside the unit.

For this project, you'll want a location with an electrical outlet, out of the way of foot traffic, and where curious pets and other animals can't get at the eggs (or break the incubator trying). A dark corner of a shed or the garage is ideal, especially where the temperature doesn't fluctuate radically over the course of the day. Keep the unit out of direct sunlight, which can quickly overheat the incubator.

There are many acceptable surfaces on which to sit the incubating eggs; this unit uses the inexpensive straw you'd find in a chicken's nest. The eggs should be placed in the straw with their pointed ends down, and you must regularly move them. Turn the eggs once or a twice a day, if possible, to ensure healthy chicks. The eggs should be placed on their sides right before they hatch.

Finally, don't be impatient. The typical incubation period for chicken eggs is twenty-one days, but that's just a guideline—some eggs will take longer. Don't be shocked if the occasional egg doesn't hatch; industrial poultry operations hatch only about eighty to ninety percent of their eggs.

WHAT YOU'LL NEED

Time: **45 minutes** | Difficulty: **Easy** | Expense: **$**

TOOLS:

Sharpie

Frameless hacksaw

Utility knife

Cordless drill and bits

MATERIALS:

Lamp socket with separate threaded rim or flange, cord, and in-line switch

Analog probe-style air conditioning thermometer (or equivalent)

Clean 5-gal. bucket with lid

Silicone sealant

Foam insulation board

Automotive cup holder

(2) #4x½" flathead screws

75-watt light

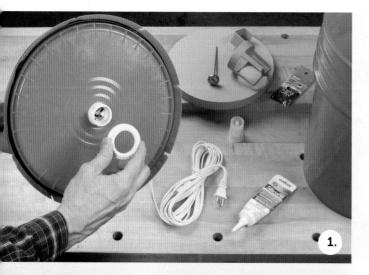

HOW YOU MAKE IT

1. Hold the lamp socket upside down, centered on the top of the bucket lid. Use the Sharpie to trace around the post of the lamp socket, then cut out the hole with the frameless hacksaw. Push the socket through the hole so that it is sitting on its rim, and screw the flange on top. Lay a bead of silicone sealant around the edge of the flange before pressing it down into position.

2. Use the bottom of the bucket to trace a circle on the insulation board. Cut inside the traced line with a utility knife. Place the foam insulation circle in the bottom of the bucket.

3. Use the utility knife or hacksaw to cut the hanging lip off the cup holder. Position the cup holder along the inside wall of the bucket, with the top about 4" below the top of the bucket. Drill 2 pilot holes through the tang of the cup holder out through the bucket. Fasten the cup holder in place by driving 2 flathead screws in from the outside through the pilot holes.

4. Opposite the cup holder and 6" from the bottom of the bucket, drill an access hole for the thermometer probe. The drill bit should be the same size as the probe or just slightly smaller. Apply a generous bead of silicone sealant around the outside of the hole, and then push the thermometer through the hole and seat the back of the dial in the sealant.

5. Drill 3 vent holes ⅜" in diameter equidistant around the outside of the bucket, about 5" from the top. To set up the incubator, spread about 2" of loose straw in the bottom of the bucket. Put a paper cup full of water in the cup holder. Plug in the light and snap the lid on the bucket. When the thermometer shows that the temperature is about 99°F, remove the top and place the eggs in the straw, pointed side down.

INCUBATING EGG AFTERCARE

Incubating eggs is a fantastic way to turn fertilized eggs into new chickens. You don't need a whole lot of specialized knowledge or the services of a veterinarian, but successfully incubating chicken eggs does take a fair amount of care and attention to details. The more TLC you lavish on incubating eggs, the more eggs that hatch in reward for your efforts. Ideally, check eggs during the process several times per day. Pay particular attention to:

• **WATER LEVEL.** The cup in the cup holder should be at least half full of water at all times. This is essential because humidity plays a key role in chick development during incubation. Chances are, you'll encounter a lot of conflicting advice about humidity. Some experts recommend starting out with around forty percent humidity for the first sixteen to eighteen days, increasing to around sixty percent for the remainder of incubation, but this incubator shoots for a constant level of humidity. Use an inexpensive hygrometer to monitor humidity in the incubator. The optimal humidity for your eggs depends on several factors—from local climate to the variations in heat in your incubator. If your hatch rate is low, you may want to add a second cup holder (although high humidity can be as bad as low).

• **TEMPERATURE.** Internal temperature should not vary by more than one or two degrees. Check the thermometer often; if the temperature starts to climb too high, turn off the light bulb. If it falls too low, use duct tape to cover the air vents until the temperature returns to normal.

• **TURNING THE EGGS.** Egg movement ensures healthy chicks. Turning the eggs several times each day is recommended. If this becomes impossible, you can purchase an aftermarket egg turner.

• **SANITATION.** Clean the incubator thoroughly between batches of eggs, with soapy water and a rinse of bleach in warm water. Also replace the straw for each new batch of eggs.

Nesting Box

When you're looking to get the most out of your chickens, there are a lot of advantages to a nesting box. A properly constructed and sited nesting box keeps hens comfortable and laying in one clean and handy place, instead of in the grass outside or other places where eggs can become contaminated or break. This makes gathering eggs much easier and ensures your yield will be as high as possible. Properly trained, the hens will always use the nesting box (see **Nesting Box Rules**, page 83). The box serves the chicken owner, as well—no more hunting for eggs!

Although it seems like a simple structure, a nesting box has certain requirements if it is going to be used as it should be. This particular box is ideal even though once you put it in place it may seem a little confining. Too much space means hens spend too much time in the box, and that usually translates to a lower number of eggs laid.

The advantages to using a five-gallon bucket go far beyond size, though. The structure is already enclosed, giving hens a sense of privacy and safety. Like any vulnerable creature, hens like to feel secure, especially when they are laying. The shape of the bucket also suits nesting perfectly. The bucket is, by its nature, adaptable and cleanable—an essential trait whenever you're building something for use by chickens.

You'll quickly know that your nesting box is successful by certain very apparent signs. First and foremost, the chickens should lay their normal number of eggs—or more if they really respond well to the box. Also, if the nesting box is reinforcing behavior as it should (and you're doing the same with how you treat the chickens), hens won't defecate in the box. They also won't peck at the eggs—a sure sign that they are not comfortable in a nesting box. More importantly, they won't lay outside of the nesting box.

If your chickens are a little reluctant to use the new nesting box, you can get them started with a little fake-out. Place a ceramic egg or even a golf ball inside the nesting box. The hen will believe the box is a safe place to lay eggs because there is already an egg there!

WHAT YOU'LL NEED

Time: 45 minutes | Difficulty: Moderate | Expense: $

TOOLS:

Straightedge

Sharpie

Vise

Jigsaw or hacksaw

Metal-cutting jigsaw blade or angle grinder

Measuring tape

Standard screwdriver

MATERIALS:

5-gal. bucket with lid

24×12×½" metal expanded sheet

Straw

(3) 1½" wood screws

½" pipe-wrap insulation

Duct tape

HOW YOU MAKE IT

1. Remove the lid from the bucket and use the straightedge and Sharpie to mark a cut line across the top of the lid, about 1" below center. Clamp the lid in a vise and use a jigsaw or hacksaw to cut the lid in two, along the cut line. Discard the larger piece.
2. Measure the depth of the bucket from the top edge to the bottom. Measure straight across the bucket about 2" in from the edge. Transfer these measurements to the expanded sheet. Cut the sheet using the jigsaw with the metal cutting blade or an angle grinder.
3. Secure the bucket in place to the wall of the coop, or to the supports under the bucket, by screwing it in place. Slide the cut expanded sheet into the bucket pushing it against the sides to secure it. Layer 2 to 3" of straw on top of the sheet. Snap the lid in place so that the cut edge is parallel to the expanded sheet.
4. Measure and cut the pipe insulation to perfectly fit the cut edge of the lid. Place the pipe insulation along the cut edge, and duct tape it in place over the bottom half of the bucket's mouth.

NESTING BOX RULES

1. NO SLEEPING ON THE JOB! Do not allow your chickens to sleep in the nesting box, because it increases the likelihood that they will defecate in the box, contaminating the eggs.

2. NO KIDS. Do not let a brooding hen raise chicks in the nesting box. The chicks will likely defecate and contaminate the box. Relocate a brooding hen to a different box or nesting area after her eggs hatch.

3. PICKUP DAILY. Harvest eggs at least once a day to protect against breakage.

4. BUILD ENOUGH HOUSING. One nesting box of the design shown here will serve three to four hens. Make sure you have enough nesting boxes for the total number of chickens.

5. COMFORT FIRST. Nesting boxes should be kept out of traffic and direct light, and should always be clean and well stocked with straw.

6. LOCK 'EM IN. Hens commonly lay early in the day, so keep them confined in the nesting box until noon.

Egg Washer

Cleaning newly harvested chicken eggs is not just a matter of having pretty eggs to put in the refrigerator or give away to friends; it's a health issue. The shells of locally grown eggs are regularly contaminated with dirt and chicken feces. In cooking, anything that's on the shell can possibly come into contact with the egg itself, so the shell should be clean.

Of course, cleaning each egg by hand—especially if you have a lot of laying hens—is a laborious prospect. That's why most chicken owners usually opt for some sort of automated cleaner. Any egg cleaner treads the fine line between a cleaning action that can effectively cleanse the surface of the shell and using so much force that you lose eggs to breakage. The best way to achieve both those goals is to use air bubbles to thoroughly scrub the eggs without breaking them.

This project uses an aerator that is similar to many compressed-air designs. The beauty of it is, you don't have to own an air compressor to run this egg cleaner. It uses the exhaust port on a shop vac. Keep in mind, though, that you need to ensure that both the vacuum chamber and filter are clean before you start blowing air into the egg cleaner.

It's important to note as well that cleaning chicken eggs correctly is essential if you want to avoid the issues that make you want to clean the eggs in the first place. Many homesteaders don't clean their eggs at all, because there is a natural protection, called bloom, on the shells. Bloom stops bacteria from migrating through the porous structure of the shell. Any cleaning typically removes the bloom, so it's important that the cleaning process not introduce bacteria or put the egg inside at risk. That's why eggs should be cleaned with specially formulated egg cleaner, and water that is at least fifteen to twenty degrees hotter than the egg itself.

Also remember that the key to successful egg washing is to ensure the eggs don't get too filthy in the first place. Collect eggs regularly—at least once a day. This ensures that they don't spend too much in the nest, where they can get dirty from chicken feet, or fouled by worse substances. Twice daily collecting is even better.

WHAT YOU'LL NEED

Time: **1 hour** | Difficulty: **Moderate** | Expense: **$$**

TOOLS:

Circular saw, jigsaw, or hacksaw

Cordless drill and bits

Clamp

Vise

MATERIALS:

½" PVC pipe (56½" section)

(8) ½" PVC 45° slip elbows

½" PVC 90° slip elbow

(3) ½" PVC slip tees

½" PVC slip ball valve

5-gal. bucket

PVC primer

PVC cement

Wire egg-collecting basket

2x1½" drain and trap connector (gray coupling with hose clamps)

1x¾" adapter, Schedule 40

HOW YOU MAKE IT

1. Cut the PVC pipe into 1½" sections to connect the elbows and the ball valve, and 1" sections to connect the inline tee (adapt as necessary if your bucket is not standard size). Cut the PVC stem 20" long and the cross piece 6½" long. Dry fit the pieces and place in the bucket to ensure they will fit.

2. Assemble the aerator by coating each connection point with PVC primer. Wait 1 minute (or the manufacturer's recommended time), then apply PVC cement and slide the pieces together. Once the aerator is completed, connect the stem and ball valve in the same way. Allow the assemblies to dry.

3. Clamp the aerator ring to a work surface and drill a random series of ⅜" holes on the top and inside faces of the PVC pieces. Secure the stem in a vise and drill random holes on one side, up about 4" from where it connects to the aerator.

4. Connect a shop-vac hose to the vacuum's exhaust valve, and use a hose clamp to connect the other end to the ball valve extension nipple. Use a step-up adapter, or multiple adaptors, as necessary to accommodate the hose diameter.

5. Fill the bucket half full of warm water; the top hole in the stem must be underwater. Add egg-cleaning soap, and carefully place the basket of eggs inside. Turn on the shop vac, and slowly open the ball valve until there is a healthy stream of bubbles surrounding the eggs. Let it run for 5 minutes.

6. Remove the eggs. Empty the bucket and rinse it out. Refill with warm water and replace the eggs. Rinse for an additional 5 minutes, then remove the eggs and dry them. Remove the aerator and hose it down. Empty the bucket it, rinse it out, and turn it upside down to dry.

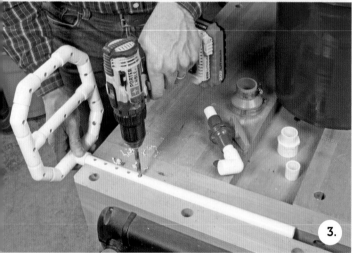

Raising pigs means feeding pigs. Feeding pigs can be a messy, time-consuming business. Make it less so with this feeder. The idea here is to use the natural curve of the five-gallon bucket to capture pig food that might have been spread out of the feeder by the pig and therefore wasted. The bucket is also large enough to hold a day's feed for an adult pig or

two (usually around six pounds), meaning less labor in filling the feeder. It is the ideal shape for a pig's snout and well-suited to different heights and maturities.

The durability of a five-gallon bucket is also ideal for a pig trough. Adorable as pigs can be, they are not exactly graceful. Any trough needs to put up with a beating at dinnertime. It also has to be stable so that hungry pigs don't overturn the feeder and waste a whole lot of pig food. The trough design here ensures stability with a PVC pipe base that is as close to indestructible as you can get. It works like a charm with the shape of the trough, making it extremely hard to tip over the unit even if you wanted to.

Unlike other common materials, such as wood and concrete, bucket plastic is not absorbent—which means that all the liquid and food meant for your pigs stays in the trough rather than being sucked up by dry material. That also accounts for fewer accumulated smells over time—always a big issue when raising pigs (especially for pig farmers who visit their charges every day). Overall, this trough is inexpensive to make, easy to put together, and will last a good long time.

Even the best feeders are only vessels. Keep in mind that pigs get the most out of a feeder into which the right food is put. Pigs, like many animals, prefer a varied diet. Ideally, you shouldn't have to give your pigs any supplements; they should get all the vitamins and nutrients from the food you provide. They also need clean water with their food—a good reason to make two of these troughs and fill the second with water.

WHAT YOU'LL NEED

Time: **45 minutes** | Difficulty: **Easy** | Expense: **$**

TOOLS:

Sharpie

Straightedge

Cordless drill and bits

Jigsaw

Small crescent wrench

MATERIALS:

5-gal. bucket with lid

Clamp or weight

(6) 2" sheetmetal screws

2½" PVC pipe (38" section)

Hot glue gun with glue or duct tape

(8) ¼" roundhead bolts and nuts, with matching rubber washers

HOW YOU MAKE IT

1. Lay the bucket on its side with the lid in place, and weight the inside or clamp the bucket to the work surface so that it won't roll when you mark the cut lines.

2. Measure 6" up from the work surface on each end of the bucket, and mark a straight horizontal line through each of those points. Extend the lines around to the sides of the bucket and connect the lines on each end.

2.

3. Drill a ½" access hole at one end and cut along the marked line with a jigsaw.

4. Drill pilot holes through the lid rims of each bucket half. Screw the buckets together end to end, through these holes.

5. Saw the PVC pipe into 9½" sections. Sit the buckets on a flat, level work surface. Butt a pipe section up to one side of one bucket half, running parallel with the bucket. Tack it in place with duct tape or dabs of hot glue. Drill two ¼" pilot holes spaced evenly apart, through the inside of the bucket and into the pipe. Bolt the pipe to the bucket with the bolts, locating the nut on the pipe side. Repeat with the three remaining pipe sections. Clean the trough and place it in the pigpen.

4.

5.

Bee Feeder

Bee feeders are a point of contention among beekeepers. Some provide their bees with supplemental food on a regular basis, while others believe that nature should take its course and that feeding bees is not only unnecessary, but wrong—and that bees given supplemental food will become dependent on the feeder rather than seeking their own food.

The fact is, whether you think the bees should find their own food or not, in times of stress—when food sources are out of season, when hives are just getting established, or when there is illness present—a supplemental food source can be a boon to the hive. It may, in certain circumstances, even be crucial to the hive's survival.

The types of bee feeders available are as varied as beekeepers' opinions. There are hive feeders that actually replace a frame inside the hive, entrance feeders that attach to the entrance of the hive, hive-top feeders, and more. The feeder in this project is one of the many external feeders that are not housed inside or on the hive, but stand right next to it. Although you give up some of the protection from robbers, such as ants, that an internal feeder affords, this feeder is simpler to maintain and could not be easier to fabricate.

It feeds the bees in narrow, under-lid cavities that should present no risk of bees drowning—a key concern with any feeder. This unit works on a simple gravity system and

hydrodynamics. A vacuum inside the bucket, created by the lid's seal, prevents overflow. As the bees feed and lower the level of syrup in the cavities, small holes release more syrup.

The formula for the syrup used in the feeder changes based on the season. Learn the specifics about different formulas in **Best Bee Food**, below. No matter what food you're putting in it, the bucket for the feeder must be clean and food grade, and have a lid with an integral gasket (to ensure an airtight seal) and a ridge below the rim that features ribs spaced regularly around the ridge.

The feeder presents a tempting food source for other insects. That's why the directions include locating it in an overturned rubber or plastic garbage can lid filled with about ½" of water.

WHAT YOU'LL NEED

Time: **20 minutes** | Difficulty: **Easy** | Expense: **$**

TOOLS:

Cordless drill and 3/32" (or similar) bit

MATERIALS:

Food-grade 5-gal. bucket and tight-fitting lid (integral gasket)

Plastic or rubber garbage can lid

HOW YOU MAKE IT

1. Remove the lid and place the bucket upside down on a work surface. Drill 2 holes in each space formed by the ribs separating the underside of the ridge.

2. Fill the bucket half full of syrup. Attach the lid and place the garbage can lid upside down next to the hive. Fill with about ½" of water. Place the feeder upside down in the center of the garbage can lid.

BEST BEE FOOD

You can't get much simpler than sugar water, but even this basic concoction spurs debate among beekeepers. Regardless of concentration, to make the sugar water (commonly known as simple syrup among bartenders and experienced beekeepers), simply combine the correct amounts of sugar and water in a large pot, bring to a simmer, and stir for four to five minutes, until the sugar completely dissolves. Allow the mixture to cool and then pour it into the feeder.

• **LIGHT CONCENTRATION.** This is made from one part sugar to two parts water. This is generally best in early spring, just as the weather starts to warm.

• **MEDIUM CONCENTRATION.** A balanced mix of one part sugar to one part water. This is a nectar-replacement food, valuable when the usual nectar sources are not producing, for whatever reason.

• **HEAVY CONCENTRATION.** This is a sugar-heavy concentration, featuring two parts sugar to one part water. This is ideal for fall or early winter in areas where plants go dormant quickly.

Do not use feeders when hive supers are in place.

3 FAMILY FUN

THERE IS A LOT OF FUN TO BE HAD WITH A FIVE-GALLON BUCKET AND A FEW SIMPLE TOOLS. AS HANDY AS THIS ALL-PURPOSE CONTAINER CAN BE, ITS ADAPTABILITY REALLY BECOMES APPARENT WHEN YOU USE IT AS THE BASIS OF FRIVOLOUS, PURE-ENJOYMENT INVENTIONS.

The fun in many of these projects lies every bit as much in the crafting of the projects as it does in the using of these wonderful creations. And this is the type of fun that can include younger members of the family. Most of the projects in this chapter can involve children; older kids may even be able to make many of these all by themselves, with minimal adult input. As a bonus, several of these projects involve some basic science concepts that can be used as a jumping-off point for educational exploration (there's no learning like fun learning!). The more you get children involved, the more of a "family project" it becomes and the more they have a feeling of accomplishment.

On top of all that, keep an open mind when working on these projects with your children, and the kids may come up with ingenious adaptations—and might even develop whole new creations.

In any case, it bears mentioning, especially where children are concerned, that you should always follow proper safety precautions. Children of any age should be closely supervised when using power tools—or even when simply near power tools that are in use. Use these projects as a chance to teach basic safety, including always cutting away from the body and using approved safety eyewear and work clothes, such as enclosed-toe shoes and gloves.

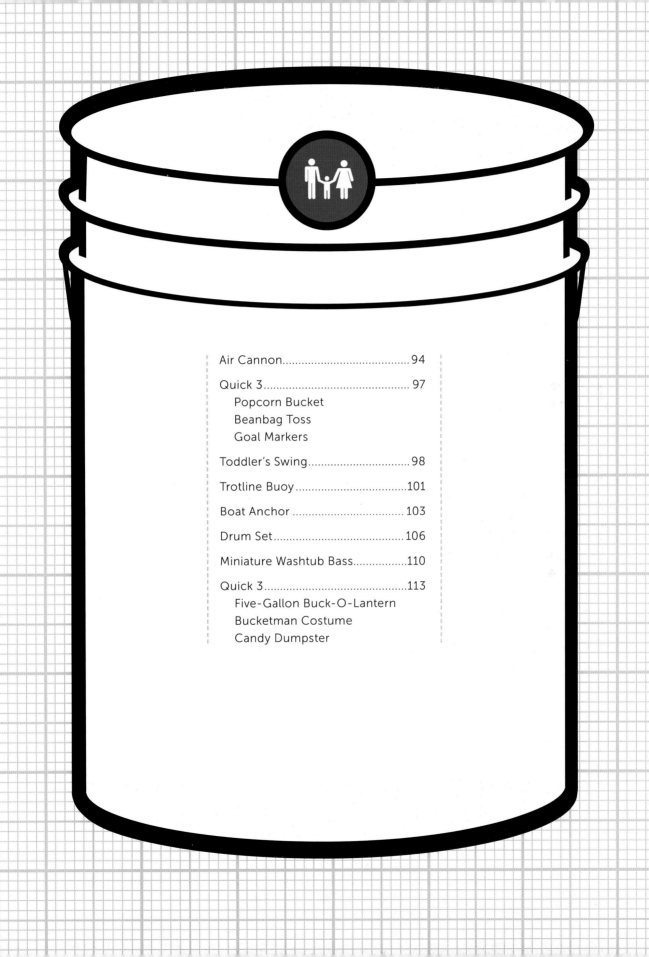

Air Cannon

This "weapon" is a lot safer than any toy gun you can bring in the house and, in action, seems to work like magic. An airtight membrane serves as the firing mechanism and a ball of air is the projectile. That may sound like watered-down fun at best, but the projectile carries very real force. Kids will be utterly amazed at the effects of the invisible-air "cannonball."

This is a fantastic opportunity to teach some simple science in action. The particular brand of physics that rules the movement of air is called fluid dynamics. The secret to an air cannon is the creation of a vortex—a swirling motion much like what you would see in water circling a drain. That action, created inside the cannon, means the air coming out of the hole carries enough energy to travel a significant distance—twenty to thirty feet!

There are all kinds of fun activities to do with an air cannon. Your youngster can collect smoke from a charcoal grill or other source and blow long-traveling smoke rings (a great way to actually see the vortex action of the air out of the cannon). The blast from the cannon can extinguish candles—the safe and sane equivalent of shooting bottles with a BB gun. Left to their own devices, children of all ages

will think up a wealth of fun games and uses for the air cannon, discovering more about the science of air in the process.

From there, it takes very little prompting for youngsters to make the connection between the effects of their ingenious toy and the immense power of a hurricane, tornado, or waterspout. That makes this toy one more spectacular leaping-off point for further investigation and research that could become a fascinating school science project and more.

The children can also be involved in making the air cannon, which can be done with a minimum of materials and very little expertise. None of the measurements need to be exact for the cannon to work, and your children can even customize the cannon by drawing designs on the outside.

WHAT YOU'LL NEED

Time: **1 hour** | Difficulty: **Moderate** | Expense: **$$**

TOOLS:

Compass

Cordless drill and ⅜" drill bit

Sharpie

Tape measure

Utility knife or scissors

Straightedge

Hole punch

8" hole saw (optional)

Frameless hacksaw (optional)

MATERIALS:

5-gal. bucket

80-grit sandpaper

1×4" threaded PVC nipple

1" threaded PVC cap

#4×½" flathead wood screws

6-mil plastic sheeting (clear preferable, but black will work)

Duct tape

Canopy tarp tie

HOW YOU MAKE IT

1. Use the compass to mark an 8" hole on the bottom of the bucket. Cut the hole out with a hole saw or drill a pilot hole and use a frameless hacksaw. Sand the cut edges of the hole smooth.

2. Screw the PVC cap onto the PVC nipple. Drill a pilot hole about 5" down from the top edge of the bucket. Drill a matching pilot hole in the center of the PVC cap, and then screw the bucket to the cap from the inside of the bucket, to create a handle.

3. Use the mouth of the bucket as a template to trace a circle in the middle of the plastic sheeting. Remove the bucket and measure out 4" from the circle all the way around, to trace a larger circle. Use the utility knife or scissors to cut out the larger circle. (Don't

2.

3.

5.

8.

worry if the children do this part and there are small imperfections—they won't affect the performance of the completed air cannon.)

4. Fold the plastic circle in half and measure and mark the exact center along the straightedge. Use the hole punch to punch a hole at this mark. (This is something that kids have plenty of experience doing, and it's a good point in the construction to get them involved.)

5. Lay a square of duct tape around the hole, to reinforce it. Feed the string end of the tarp tie through the hole until the ball end is snug against the other side. Tie a slip knot in the string of the tie, and snug it against the side opposite the ball. Cut the loop of the tarp tie, so that there are 2 loose ends.

6. Use a straightedge to mark 2 points on the bottom rim of the bucket directly across from each other. Measure up the sides of the bucket 2" from each point and make a mark. Drill ⅜" holes at these marks.

7. Center the plastic sheeting circle over the mouth of the bucket with the ball of the tarp tie on the outside. Pull the edges of the plastic almost taut (there needs to be some play for the cannon to work), and use small strips of duct tape on 4 sides to hold it in place (this is much easier with a helper and a chance to make a child feel involved in the process). Lay a strip of duct tape around the edge of the plastic, all the way around, to secure it to the bucket. Make sure the duct tape has been smoothed down and is holding firmly to the bucket.

8. Pull one of the loose ends of the tarp tie through one of the holes in the bucket. Tie a knot at the end, on the outside of the bucket. Repeat with the other loose end and the other hole, and you're ready to test the cannon.

QUICK 3

Three Quick and Useful Bucket Options

Family fun with buckets doesn't necessarily require involved fabrication or lots of time. Here are three that provide plenty of entertainment with minimal expense and work.

1.
Popcorn Bucket

This gift container holds treats that are easy to make, but that will delight adults and children alike. Measure the inside radius at the top and bottom of the bucket and transfer those measurements to two fourteen-inch cardboard squares. Cut them into tapered rectangles, following the measurements. Cut a slot along the length of one piece from the bottom to the middle, and on the second piece from the middle to the top. Now slide the pieces together along the slots so that they form a cross. Slide this down into a clean five-gallon bucket to make four separate compartments. Decorate the bucket and fill each quarter with a different flavor of popcorn. Or substitute hard candies or other treats.

2.
Beanbag Toss

Create a simple-but-engaging yard game for warm-weather fun by filling six plastic sandwich bags half full of dried beans. Wrap colored duct tape tightly around each bag. Use a hole saw to drill a hole just slightly larger than one of these beanbags in the center of a bucket's lid. Use a Sharpie to draw a second circle around the hole in the lid, and mark the lid's rim (label with scores for the game). To play, prop up the bucket so that the top is angled toward the players. Each player gets three throws; high score wins.

3.
Goal Markers

Bright bucket goal markers are perfect for quick neighborhood soccer games. Wrap blue painter's tape around a bucket in a spiral pattern and then paint the bucket in neon orange or green, or "caution" yellow. Remove the tape, turn the bucket upside down, and you have an easy-to-see marker. For a weighted marker that won't topple, fill the bucket half full of sand or dirt.

FAMILY FUN

Toddler's Swing

There's something magical about a swing. Whether it's a tire hung from the biggest branch in a backyard walnut tree, or a more formal part of a swing set, a swing seems to invite hypnotic relaxation. Pull yourself back and then let go and yield to a simple pendulum motion. The motion is intuitive and never grows old. Most people, whether they'll admit it or not, won't ever grow out of the desire to have a leisurely swing.

But some people are just too young for a simple plank suspended by two chains or ropes. The irony is, those youngest members of the family are the ones that delight the most in a little free swing time. Well, now they need not be left out of the fun.

The toddler's swing described in this project is meant to support a youngster who hasn't yet developed the muscular coordination and balance necessary to use a plank-type swing. The design is similar to the rubber and burlap swings used on playgrounds across America.

The swing is hung by chains, which are more durable, easier to connect, and easier to adjust than ropes. They allow the swing to be suspended from the sturdy six-by-six crossbeams on an existing play system, the metal support of a swing set (remove one of the big child swings), or even the level branch of a large healthy tree.

Depending on their size, children from about one year old up to three years old should reliably be able to use this swing. You'll need to exercise some judgment; the child should be able to sit comfortably without slumping, and should not be so big that they are cramped for space. If any part of the swing is clearly too tight, it may be time to move to a big-kid version.

In any case, test whatever support you're thinking of using to ensure it will hold the weight of an adult—that will guarantee the little person is safe and sound for all the swinging he or she can handle.

WHAT YOU'LL NEED

Time: **45 minutes** | Difficulty: **Moderate** | Expense: **$$**

TOOLS:

Sharpie

Cordless drill
and 2" hole saw

Frameless hacksaw

MATERIALS:

Butcher or other kraft paper

Clean 5-gal. bucket
without handle

100-grit sandpaper

(2) ½" zinc-plated steel
spring links

(2) foam garden kneeling
pads (or scrap pieces of
dense foam)

Spray foam adhesive

(3) sections ⁵⁄₁₆" zinc-plated
chain (length as needed
based on swing height)

HOW YOU MAKE IT

1. Draw the swing design cut lines on a large section of butcher paper. The two leg holes will be roughly 6" ovals, overlapping the side and bottom of the bucket; the front edge should be about 6" from the bottom and should gently curve up to the back at about the midpoint of the bucket on each side. The back should be cut to right below the bottom ridge line on the bucket. (Exact dimensions can be adjusted depending on the size of the child and the bucket you're using.) Wrap the butcher paper around the bucket and secure it with tape. Use the Sharpie to trace the cut lines.

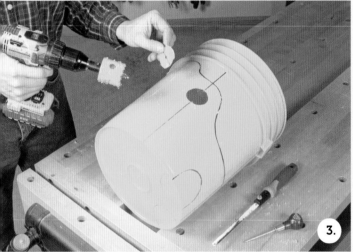

2. Remove the paper and make sure the cut lines are apparent all the way around the bucket. The entire design should fall below the bottom ridge at the top of the bucket.

3. Measure and mark the center of the chain holes at the midpoint on each side, as shown in the photo. (There should be at least 2" of plastic between the top of the hole and the top edge of the swing arm). Cut out the chain holes with the hole saw. Use the frameless hacksaw to cut out the seat shape. Sand all the edges smooth.

4. Cut the foam for a seat and back. Coat it with adhesive and clamp it in place. Attach two equal lengths of chain to the swing handles with the spring links. Connect the chains to a cross brace or loop them over a sturdy branch and spring clip each to itself. Slip your toddler into the swing to check fit, and adjust as necessary to make the swing more comfortable.

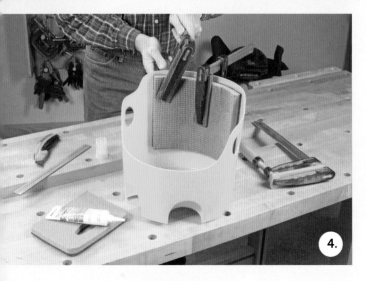

Trotline Buoy

Trotlines are heavy fishing lines with lighter baited lines (called snoods) running off them. The idea is to pull in as many water edibles as possible with the least amount of time and effort. Commercial crab and catfish fisherman use trotlines to haul in big catches. To work correctly, the trotline must be held up by buoys (although they are also commonly strung all the way across a body of water). A word to the wise—although a trotline is a good way to bring in a boat full of fish, they are illegal in some states and some areas of other states. Check your local regulations before using the buoy in this project for your own personal trotline.

Although five-gallon buckets make ideal, highly visible trotline buoys, the buoys themselves can also be adapted for many other purposes. You can use them for marking shallow depths in a small lake or anywhere else recreational boating is common. They can be used to mark off certain areas, such as coves. And they can even be used to mark out a course for a swimming competition. Wherever there is a body of water, a buoy can come in mighty handy.

It's important that the five-gallon buckets you use for this project have integral lid gaskets. These gaskets will help ensure against leakage so that your buoy keeps on floating regardless of how battered or buffeted it might get. If you absolutely can't find buckets with intact lid gaskets, lay a bead of silicone sealant in the lid channel before securing the lid on top. Then let it dry before using the buoys.

You should also be aware that sun and water exposure will eventually cause the plastic in the buckets to become brittle and will fade the color (colorful, easily seen buoys or floats are essential to running a trotline). You can paint the buckets with heavy-duty exterior paint meant for plastics (see **Painting Your Bucket** on page 56) or pick a brightly colored bucket to start with. The key to the success of any buoy is its ability to be seen from a distance. Expect to replace the bucket buoys every one to two years under normal use.

WHAT YOU'LL NEED

Time: **30 minutes** | Difficulty: **Easy** | Expense: **$**

TOOLS:

Cordless drill and bits

(2) crescent wrenches

MATERIALS:

(4) ⅜" stainless steel nuts

Yellow or orange 5-gal. bucket with lid (with integral gasket)

⅜×6" stainless steel threaded rod

(2) ⅜×5" stainless steel eyebolts

(2) ⅜×1⅛" stainless steel rod coupling nuts

Silicone sealant

Expanding foam insulation

Flange nuts and rubber washers (optional)

HOW YOU MAKE IT

1. Remove the bucket's handle. Drill through the bucket handle holes on each side, using a ⅜" bit.

2. Push the eyebolts in from each side so that the eyes are on the outside of the bucket. Screw a plain nut onto each end of the threaded rod and the end of each eyebolt, screwing them on far enough to leave room to attach the coupling nut. Connect the two eyebolts through the interior of the bucket, with the threaded rod, using the coupling nuts. Use the crescent wrenches to tighten until the assembly is snug, and then secure the tightening nuts with the plain nuts on either side.

3. Lay a bead of silicone sealant around and in the eyebolt hole, on the inside and outside of the bucket. Let the sealant dry completely.

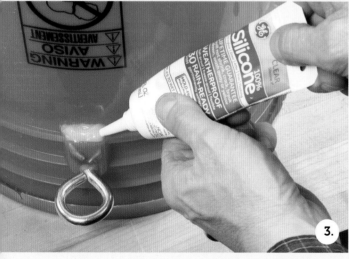

4. Turn the bucket upside down and fill the cavity between where the handle went through the ridge and the corresponding hole in the side of the bucket with silicone sealant. (If your bucket is the type with a solid plastic handle block molded into the side of the bucket, you won't need to do this.)

5. Repeat with additional buckets as needed for your trotline. Make sure the lids are secured on top prior to using the buoys. Use swivel snaps to connect the trotlines to the buoys. If there is any leakage through the handle holes, disassemble the eyebolt construction and use flange nuts with rubber washers on each side of both eyebolts.

Boat Anchor

A boat anchor comes in mighty handy when you want to park a boat near a favorite fishing hole with a strong current underneath. It's also essential at other times when you want to come ashore but for one reason or another you cannot beach your boat. This anchor can also be used to secure a swimming platform or even a buoy to mark shallow depth or to run a trotline.

In boating, there are different types of anchors for different situations, and this one is a cross between two common general-purpose anchors: plow and mushroom. However, it's wise to keep in mind that no one anchor will serve every purpose, and it's always better to be safe than sorry. That's why many boaters use at least two anchors. You can modify this design to suit your own boating purposes. In any case, it is intended only for small, light boats. Larger boats require a heavier metal anchor.

One of the big benefits of this particular anchor is that it's self-storing. When not in use, you simply coil the anchor chain or rope

inside the bucket itself. There's even extra room to hold other gear and tackle as need be. This particular project is also very forgiving. You don't need to be perfectly exact with the measurements, cutting, or anything else in the fabrication of the anchor. The finished product will be durable and useful on the water, and can even be painted if you want to spruce up the look of your boat (although water has a way of wearing everything down to its basic nature).

Regardless of where you set the anchor down, follow basic nautical practices, even in a small boat. Anchor in an area out of the way of boating traffic. You'll also want to set down in a place out of the wind. It's best to choose a location with optimum "swing room" because the boat may be subject to movement with the tides or wind, and you don't want to hit anything or be grounded. The most important rule is never to anchor from the stern, because this could cause the boat to capsize.

WHAT YOU'LL NEED

Time: **45 minutes** | Difficulty: **Easy** | Expense: **$**

TOOLS:

Vise

Hacksaw

Cordless drill and bits

2" hole saw

MATERIALS:

2" PVC pipe (36" section)

5-gal. bucket

Duct tape

3⁄8×4" eyebolt

(2) 3⁄8" nuts

3⁄8" fender washer

Bag of quick-dry cement

1⁄4" zinc-plated chain (or nylon anchor rope)

4" marine snap hook

1.

HOW YOU MAKE IT

1. Secure the PVC pipe in a vise. Using the hacksaw, saw it into 3 sections of 12" with sharply angled cuts on one end.

2. Use the hole saw to drill 3 holes in the bucket's bottom, spaced equally apart in a triangle. Push the pipes through the holes, spike ends downward. About 8" of each pipe should project out of the bottom. Tape the pipes in place, skewed at angles. Tape completely around the pipe holes to prevent wet concrete from leaking out.

3. Hold the eyebolt between the tops of the anchor spikes inside the bucket. Thread a nut halfway up the eyebolt threads, slip on the fender washer, and then thread the second under the washer. Tighten the nuts to secure the washer.

4. Fill the bucket halfway with the quick-dry cement. Add water and use a scrap stick to poke the concrete to eliminate air holes. Add more water as necessary if the concrete is too dry. Strike the side of the bucket several times to eliminate air holes and allow the concrete to dry.

5. Drill ¼" drainage holes all the way around the bucket, just above the top of the concrete. Affix the anchor chain or rope to the anchor eyebolt with the snap hook.

2.

3.

5.

Drum Set

Few musical instruments capture and nurture a child's enthusiasm and musical interest like a drum set. Banging the drums is not only a great way for young musicians to learn about rhythm, it's also a wonderful outlet for youthful energy. Drums are a great introduction to any kind of music because the basics are some of the easiest to learn among musical instruments.

Drums are also forgiving, something this drum set trades on. Unlike woodwinds or a guitar, drums don't really need to be tuned. As music teachers have long known, just about anything you can bang can be a drum.

The downside is that high-quality drum sets can carry a big price tag. Given how fleeting children's fascinations are, many parents are leery of footing the bill for a pricey instrument that might wind up gathering dust in a month's time. No such worries with this homemade set. It includes almost all the basic parts of a professional drum kit, crafted out of found materials. You can even use rough buckets that wouldn't be appropriate for more exacting projects or those that involve food or drink.

The beauty of this particular project is that kids will have almost as much fun participating

in the fabrication process as they will in using the end product.

The set includes a bass drum, standing or "floor" tom, rack-mounted toms, and a snare. It does not include basic cymbals or a high-hat, but these can be homemade as well with a little ingenuity and a cut-down garbage can lid or recycled pot lid mounted on a tripod, similar to the one described for the snare drum. You'll have to buy a pedal for the bass drum, or discuss with your young drummer how the pedal functions and put your heads together to create one from scratch. You'll find instructions for a usable stool on page 54.

WHAT YOU'LL NEED

Time: **2 hours** | Difficulty: **Moderate** | Expense: **$$**

TOOLS:

Measuring tape

Cordless drill and bits

Jigsaw or hacksaw

Utility knife

Scissors

MATERIALS:

2x6x12" pine or cedar

(4) 5-gal. buckets, handles removed

(6) 5-gal. bucket lids

#4x1" flathead wood screws

#4x2" machine screws and nuts

Sharpie

(2) 1½" threaded PVC caps

Duct tape

(4) ⅜x2" bolts and nuts

(8) ⅜" fender washers

1½x4" threaded PVC nipple

1½x2" threaded PVC nipple

1½" threaded PVC 30° elbow

½x36" threaded rod with 6 nuts

3" PVC pipe (16" section)

2x12" threaded nipple

Vise

Heavy-duty aluminum foil

Container of BBs

Strong plastic adhesive

2" threaded PVC flange

2" threaded nipple and coupling

HOW YOU MAKE IT

1. Lay the 2x6 board on a work surface. Remove the lid and lay the bucket on its side on top of the board, so that the lowest top ridge of the bucket is off the board. Drill 3 wood screws from inside the bucket into the board, spacing them evenly along the length.

2. Measure and mark 2 points about 7" from the front end (lid end) of this bucket, on the inside—approximately 1" on either side of the bucket's top centerline. Place 1½" threaded PVC caps at these points, top down, and drill pilot holes through the caps and bucket. Move the caps to the outside of the bucket and fasten them with machine screws and nuts.

1.

2.

3. Cut a 5-gal. bucket exactly in half crosswise, using a jigsaw or a hacksaw. Use the bottom (cut side) of the upper half as a template to draw a circle on a scrap 5-gal. bucket lid. Cut out the circle with a utility knife, cutting slightly outside the marked line (or use the hacksaw or jigsaw). Use the bottom (cut side) of the lower, smaller half as a template to draw a circle on a second scrap 5-gal. bucket lid. Cut it out as you did the larger circle. Drill 2 holes in each tom for the bolts and nuts to hold the flange to the toms. Attach the flanges with 2" bolts, secured with a nut, with a fender washer on each side.

4. Duct tape the respective circles in place on the two halves to create small and large rack-mounted toms.

3.

5. Assemble the tom posts by screwing a 1½×4" threaded PVC nipple into the cap on the bass drum bucket. Screw a 1½" threaded PVC 30° elbow onto the opposite end of the nipple, and attach the tom to the elbow with a 1½×2" threaded PVC nipple. Repeat with the opposite tom. Adjust the position by tightening or loosening the elbows. Note: You can adjust how high the mounted tom-toms sit by substituting different lengths of threaded PVC nipples.

6. Cut the PVC pipe in half so that you have 2 sections 8" long. Drilling from the underside of the bucket lid, screw these sections side by side on the top of the lid. Snap the lid onto the bucket and turn it over to complete the floor tom.

4.

7. Fabricate the snare tripod by cutting the threaded rod into 3 12" sections with a hacksaw. Secure a 2×12" threaded nipple in a vise and drill 3 sets of holes evenly spaced around one end, drilling all the way through the pipe at a 45° angle.

8. Stick each of the threaded rod sections through a set of holes to create tripod feet. Secure each section with a nut on each side of the pipe. The nuts allow you to adjust the feet.

9. To make the snare, cut the top of a 5-gal. bucket, 4 inches down from the top of the lid. Use the cut bottom of this top section as a template to trace a circle on the top of a scrap 5-gal. bucket lid. Cut out the circle. This will be the top surface of the snare drum.

10. Layer several sheets of the aluminum foil on a work surface. Use the circle you cut for the snare drum top as a template to draw a circle on the foil. Cut out the foil circle with a scissors, adding about 1" all the way around. Spread the BBs evenly across the foil circle. Lay the cut circle for the snare-top surface centered on top of the BBs and foil. Crimp the foil up around the edges of the plastic circle.

11. Set the plastic-and-foil snare surface over the bottom of the top section you cut from the bucket. Duct tape the circle in place. Use the plastic adhesive to fasten the 2" threaded PVC flange onto the bottom (lid side) of the drum. Center the flange on the surface.

12. Attach the stand to the drum by screwing the 2" threaded nipple into the flange and into a coupling on the other end. Screw the open end of the coupling onto the tripod. You can adjust the stand height by using longer nipples or a combination of threaded nipples and couplings. Adjust the tilt angle of the drum by moving the nuts on the feet.

13. Set up the drum set to accommodate the drummer. This may entail changing the positions of the mounted tom-toms or snare drum, or weighting down the bass drum so that it doesn't move when the pedal hits the drum. You can also raise the floor tom by placing it on a small square of plywood—and you can even screw the plywood to the PVC pipe feet on the drum.

6.

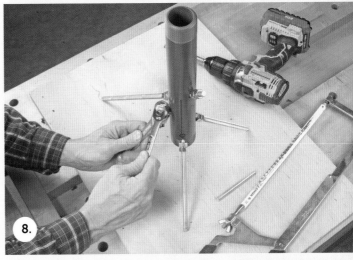

8.

10.

Not necessarily the most precise of musical instruments, the washtub bass is certainly one of the most fun to play. Also known as the gutbucket, the washtub bass comes in many different forms. It really is the classic do-it-yourself musical instrument. Some include four strings with tuning pegs, while many homemade versions use only a single string. The use of this unique and rustic instrument dates back to the early 1900s or even before, depending on whom you ask. But the roots are most definitely in rough-and-tumble country music. These days, the instrument is a standard in rockabilly bands and backwoods jams everywhere. The single string is alternately stretched and loosened to play different notes, and provides an amazingly wide range of sounds for a simple instrument.

The playing action of any washtub bass requires a moveable post. The project here uses a replacement shovel handle, but you could easily substitute an old broom handle or even a length of PVC pipe. And, of course, the project substitutes a five-gallon bucket for a

AMPLIFY THAT NOISE!

Once you get good enough to play the bass for real, you might want to amplify all those good sounds. It's easy enough to do. Drill a hole in the side of the bucket and run the cord of a microphone through the hole. Position the microphone on a microphone stand centered under the bucket (don't use the flange—the bucket should sit flat on the floor). Plug the microphone into an amplifier and boom! You're ready to rock with an electric bass.

metal washtub. The sound is somewhat different, and the size is more modest than a full-scale version, but it works fine for youngsters who are just getting their feet wet and want to experiment with making sounds.

As with all washtub basses, the idea is to use found objects, representing no cost, to make a durable and playable instrument. Making the bass requires very little time, expertise, or expense. It can be a learning experience for budding musicians in developing homemade instruments for their own musical needs. But in reality, it can be just as interesting for adult music lovers who want to try their hand at making a traditional, folksy, homemade instrument.

The bass marries best with similar rural instruments, such as a washboard, acoustic guitar or banjo, and harmonica or mouth harp. All of these are fairly inexpensive in and of themselves, so it's relatively cheap and easy to help your child put together a jug band of his or her very own.

WHAT YOU'LL NEED

Time: **30 minutes** | Difficulty: **Easy** | Expense: **$**

TOOLS:

Hacksaw

Cordless drill and bits

Jigsaw

Crescent wrench

Drilling jig (optional)

MATERIALS:

5 gal. bucket with lid, handle removed

¼" eyebolt, with 2 flange washers, 2 lock washers, and 2 nuts

48" hardwood tool handle

Sharpie

Vise

2" PVC flange

³⁄₁₆" cotton clothesline or equivalent (such as parachute cord)

⅛" wire clamp

Rubber gasket

Rubber cement

80-grit sandpaper (optional)

HOW YOU MAKE IT

1. Using the jigsaw, cut the bucket crosswise at a point 8" from the bottom. Drill a ¼" hole in the center of the bucket bottom. Secure the eyebolt in the hole, with a flange washer, lock washer, and nut on each side (in that order). If you'll be using the bass on a floor that can be scratched, sand the bottom edge of the bass.

2. With the bucket bottom facing up, rest one end of the tool handle on the bucket edge. Position it so that about ¾ of the handle diameter is over the bottom, with the ¼ hanging off the edge. Mark the stick and the bucket bottom with cut marks representing this position.

3. With the jigsaw, cut a recess in the tool handle end and a matching hole in the bucket. Dry fit the handle into the hole, and make sure that the handle will move freely without popping out of the hole. Adjust the hole or the handle-end cut as necessary.

4. Clamp the handle in a vise and drill a ⅛" hole through the handle, about 5" down from the top. A jig will be a big help in drilling the hole.

5. Set the cut edge of the bucket on top of the PVC flange and mark the position of the edge. Remove the flange and cut a slot along the mark with a hacksaw. Run one end of the clothesline through the tool handle hole and tie a knot on the end.

6. Loop the other end of the clothesline through the eyebolt, and secure it with the wire clamp, using the crescent wrench to tighten it down. The line should be fairly taut with the stick held straight up and down. Adjust the clamp to adjust the line's slackness. Cut the extra tail of clothesline, leaving 2 to 3" to allow for adjustments, as necessary.

7. To play the bass, set the edge opposite the handle on the flange so that it is slightly raised. Put your "weak" foot behind the bucket to stop it from slipping, and place your other foot on the top of the bucket to hold it down. Put the stick in the hole and hold it so that the string is taut. Strike the string with your finger to make a note, and pull back on the stick to make the notes go higher.

QUICK 3

Three Quick and Entertaining Halloween Bucket Options

The five-gallon bucket really comes into its own when holiday celebrations roll around. These projects are all about pure, silly, kid-pleasing fun. They might not be practical, but don't be surprised when the neighbors start copying your designs!

1.
Five-Gallon Buck-O-Lantern

Why kill an innocent pumpkin when you can recycle a battered old five-gallon bucket for your front porch horror show? This works best if the bucket is already orange, although it's easy enough to paint a bucket orange. Either way, draw your favorite jack-o'-lantern face on one side of the bucket. Drill pilot holes and cut out the design with a frameless hacksaw or Dremel tool. Put a tea light inside, or use a more traditional candle, but be sure to put it on a saucer, to catch melting wax, and drill vent holes in the lid.

2.
Bucketman Costume

The Halloween party starts in fifteen minutes and you don't have a costume. Are you going to cop out and go as yourself yet again? Heck no! Grab a clean five-gallon bucket, flip it upside down, and draw two large eyeholes the same distance from the top edge of the bucket as your eyes are from your shoulder. Add a mouth hole (or cut scarier mouth slots, like a hockey mask). Cut out the holes and slots with a utility knife and your costume is complete. Want to go a little further? Get out the felt-tip pens or paints and create a design like a Mexican wrestling mask.

3.
Candy Dumpster

Let's face it: the whole goal behind trick-or-treating is to collect as much candy as humanly possible. The weak link in the chain has always been the candy bag or the pillowcase. They just can't hold enough, so your child (or you, as the case may be) has to waste precious time heading home for a new candy carrier. Enter the candy dumpster. It has a magnificent five-gallon capacity and easy-to-use rollers that make hitting all the houses in the neighborhood before bedtime a breeze. Simply screw three to four swivel casters onto the bottom of a five-gallon bucket. If you're feeling ambitious, decorate the bucket to match your costume, and then head out to collect pounds and pounds of those free sweet treats.

YARD AND GARDEN INNOVATIONS

4

IT DOESN'T MATTER WHETHER YOU'RE GROWING A SMALL FARM'S WORTH OF VEGETABLES OR CREATING A TO-DIE-FOR ORNAMENTAL GARDEN, ANY HOME LANDSCAPE TAKES CARE AND MAINTENANCE. THE MANY TASKS THAT NEED DOING IN A YARD OR GARDEN ARE MADE EASIER AND LESS EXPENSIVE WITH THE HELP OF A FIVE-GALLON BUCKET.

Helping food crops grow their best is a natural for the ubiquitous bucket. Practical applications are where the adaptability of this unique container really becomes apparent. The bucket serves as an excellent planter, composter, watering device, and much more. In short, if you want the very best yield out of what you grow, turn to the bucket.

That's not to say that a properly dressed-up bucket can't serve more aesthetic purposes. You'll find one just as valuable as a garden fountain as it is a home for fast-growing tomatoes. It can even straddle that gray area between functional and beautiful, when used for a miniature spiral herb garden or a strawberry pot. Really, there's just no bad place in the yard for a five-gallon bucket. So, as the calendar marches on toward spring, it might be a great time to stock up on your collection of these plastic wonders.

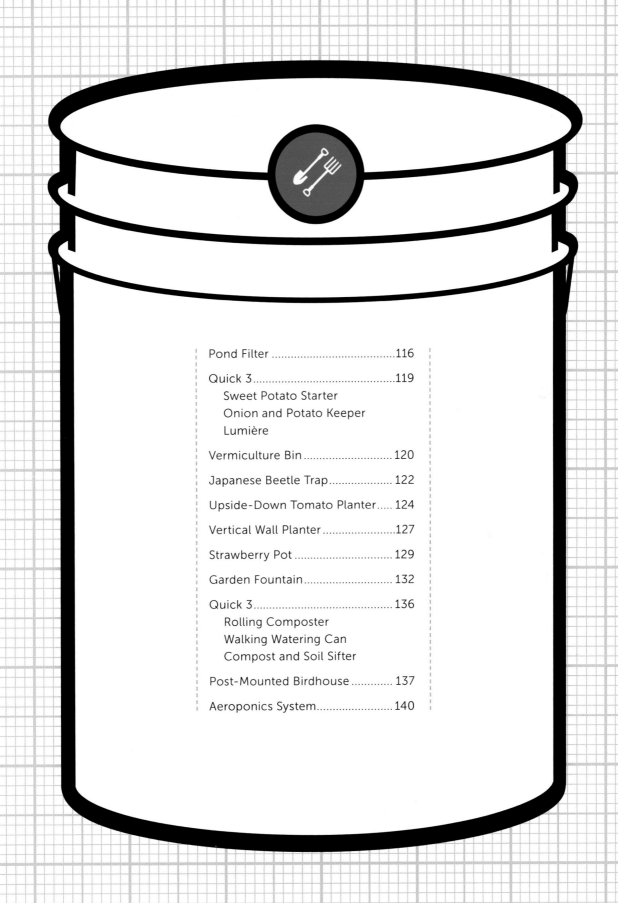

Even a small pond can transform a backyard into an oasis. It can provide a home for fascinating fish and an ecosystem for beneficial insects, allow you to grow waterborne flowers, and provide a focus to the entire landscape design. However, unless you want an algae-ridden swamp in place of that showcase outdoor feature, you must have a working and efficient pump filtering the water. The problem is, pond filters can run hundreds of dollars, and many fail after a season or two. They are

hard-working appliances that need to be durable and do their job well.

You don't need to break the bank to have a serviceable pond filter that does the job over the long haul. Instead, you can make a much more modest investment in a five-gallon bucket filter, throw in a little bit of time and labor, and have a unit that works just as well as a model you would find at retail. Whether you build or buy the filter, the principle is fairly rudimentary: run pond water through a series of increasingly finer mesh filters, and over what are known as bio balls (finned balls that help increase the amount of beneficial bacteria in the water), and then return the cleaned water to the pond at a rate that ensures the water doesn't overwhelm the filter itself.

The filter in this project can efficiently clean a pond of three hundred to five hundred gallons. If you have a significantly larger pond, you can certainly build a second filter and split the outlet line from the pump that draws water from the pond. No matter what type of filter cleans your pond water, you need to regularly clean both the pump and filters (two times per season at least) and monitor the quality of the pond water. This is especially true if you have aquatic life in the pond. It is also a good idea for aesthetic reasons to build a small wooden structure or stone enclosure to hide your five-gallon bucket filter.

WHAT YOU'LL NEED

Time: **1 hour** | Difficulty: **Moderate** | Expense: **$$$**

TOOLS:

Sharpie

Cordless drill and bits, including 1" spade bit

1½" hole saw

Jigsaw

Slip groove pliers

MATERIALS:

5-gal. bucket with lid

1½" threaded PVC bulkhead fitting, nut, and rubber washer

1" threaded PVC bulkhead fitting and nut

1x8" threaded PVC nipple

1" threaded PVC elbow

Vise

(2) 1x2" threaded PVC nipples

1½" PVC tee

1½x8" PVC nipple

(2) 1½" PVC coupling

Pond filter pads, coarse, medium, and fine

Bio balls (or use plastic pot scrubbers from the dollar store)

HOW YOU MAKE IT

1. Mark a 1½" hole at the base of the bucket, about 1" up from the bottom, using the threaded post of the larger bulkhead fitting as a template. Drill out the hole with a 1½" hole saw.

2. Mark a 1" hole on the lid at one edge (inside the rim). Use the threaded post of the smaller bulkhead fitting as a template. Drill out the hole with the spade bit.

3. Screw one end of the 1x8" PVC nipple into a threaded PVC elbow. Secure the elbow in a vise with the open leg pointed downward and the pipe parallel to the floor. Drill a ¼" hole every inch along the length of the pipe.

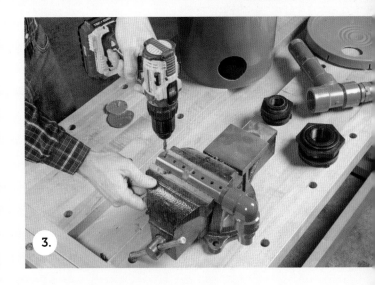

3.

4.

4. Slide the threaded post of the smaller bulkhead fitting through the lid hole from the top. Secure it with the nut, tightening hand-tight. Screw 1 of the threaded 2" PVC nipples into the free arm of the elbow and screw the other end into the bulkhead fitting in the underside of the lid. Make sure that the elbow is oriented correctly so that the inlet PVC nipple will fit inside the bucket.

5. Screw 1 end of the threaded coupling into 1 of the end legs (not the middle post) of the tee, and screw the 1½×8" PVC nipple into the middle opening.

6.

6. Slide a rubber washer over the threaded post of the larger bulkhead fitting, then slide the post through the bottom bucket hole from the inside. Thread a rubber washer and nut over the post and screw the fitting onto the end of the threaded nipple that has been secured into the tee leg. Tighten down the rubber washer and nut, hand-tight, and then a quarter turn more with the slip groove pliers.

7. Use the bottom of the bucket as a template to cut the filter media pads into 3 separate circles. Fill the bottom third of the bucket with bio balls (or scrubbers). Push each layer of filter media down into the bucket, cutting a hole for the overflow pipe sticking up from the lower fitting—fine filter first, then medium, and coarse on top.

7.

8. Screw the second 1½" PVC coupling into the outside of the bottom bulkhead fitting. Screw the second threaded 2" nipple onto the lid bulkhead fitting.

9. Attach the pump hose to the lid coupling and secure the lid on the bucket. Turn the pump on and ensure that the water is flowing correctly and that the pump isn't backing up. The pond water should begin to clear within a few days.

QUICK 3

Three Quick and Quirky Yard and Garden Bucket Options

A five-gallon bucket garden won't provide you with a complete and varied harvest, but the bucket can be useful for starting or storing a more limited crop. You can also use the bucket to brighten the night garden—it's just a matter of adopting an old idea to a new format.

1.
Sweet Potato Starter

Sweet potatoes are delicious, full of fiber, and packed with nutrients. Growing your own adds the satisfaction of free produce. To start, you'll need slips from a seed potato. Drill eighth-inch drain holes across the bottom of a five-gallon bucket and fill the bucket with soil the dampness of a wrung-out sponge. Plant an organic sweet potato with the top one-third exposed. Keep the soil moist and the seed potato will sprout "slips"—stems that can each be planted to produce a new potato.

2.
Onion and Potato Keeper

When your onion and potato harvest comes in, or when you buy in bulk, you don't want any of that usable produce to go to waste. The way to keep it, and keep it handy, is a five-gallon-bucket cold cellar. Use a two-inch hole saw to drill several holes around the body of the bucket, and cut a large access opening at the bottom of the bucket on one side (you should be able to pass a large potato through it). Then stack the potatoes in layers separated by layers of straw. Do the same with the onions and keep the bucket in a cool, dark space, such as a closet or garage.

3.
Lumière

Lumière is just a fancy word for a beautiful garden lantern that doesn't require electricity. Lumières are popular in gardens around the world and are often used to mark special occasions outdoors. They are usually small, but this lumière can be a wonderful centerpiece to an outdoor party. Choose a simple surface design involving small openings in the surface of the bucket—the easiest is just to drill a constellation of quarter-inch or smaller holes. Add about two to three inches of sand in the bucket and place a large pillar candle in the center. Light the candle for a lovely, mood-setting glow that will bring the surface design to life.

Vermiculture Bin

Worms are some of the most beneficial creatures in any garden. They are unrivaled for breaking up hard, compacted soil, allowing water, air, and nutrients to reach plant roots. They also break down organic matter—both by burrowing and eating—improving soil texture in the process. But the best thing they do for your garden is poop. Worm poop—more elegantly known as castings—is some of the richest natural fertilizer you can use.

The trouble with letting worms roam free in your yard is that they are just as likely to make their way to your neighbor's yard or destinations unknown, taking all those great benefits with them. Worms are fickle that way.

The secret to exploiting worms is to create your own worm farm. Confining them to a vermiculture bin like the one described here means that you can make the maximum amount of castings in the shortest amount of time. Under ideal circumstances, a worm will eat its way through its own weight in food in a single day. One of those circumstances is having the right type of food. Feed your worms quality kitchen scraps (no meat, no dairy, and no refined or processed foods, such as hamburger buns). Worms can also eat their way through paper products, but avoid anything with ink or that is coated, and ideally use acid-free paper.

Worms can drown, so you don't want the pile in your buckets to be sopping wet. However, you also don't want it to be bone dry. If it's about the same as a wrung-out sponge, you'll be fine. And not just any worms will do; buy red wiggler composting worms, widely available online or in bait stores.

The bottom bucket in this vermiculture bin collects naturally occurring fluids known as leachate, produced in the composting process as worms eat their way through organic matter. Both the compost itself and the resulting fluid in the bottom bucket are like manna to your plants—no matter what types of plants they are. (Even lawns love this type of fertilizer.) Regularly empty the lowest bucket when it's full of almost-black, fine-particulate castings. Use the castings just as you would fertilizer or compost.

TOOLS:

Cordless drill and bits

MATERIALS:

(3) 5-gal. buckets, dark-colored, if possible

Bucket lid

(2) 6"-tall planting pots (sturdy plastic or terra cotta)

HOW YOU MAKE IT

1. Drill ⅜" holes in a random pattern all over the bottoms of 2 of the buckets. Drill ⅛" holes all around the top of each of those 2 buckets, in a ring under the bottom ridge at the top.

2. Drill ⅛" holes in a random pattern across the lid. If you're using plastic pots, drill ⅛" holes all around the outside of the pots.

3. Set up the bins by placing the undrilled bucket on the bottom. Put a plastic pot upside down in the bottom of the bucket (centered), then slide 1 of the drilled buckets into the bottom bucket. Place the other plastic pot upside down in the bottom of the drilled bucket.

4. Fill the bucket with shredded paper—ideally paper with no printing on it. You can also use coarse sawdust from untreated wood. Add ½ pound of worms and a food source (coffee grounds and crushed eggshells are excellent starter foods for worms).

5. Slide the second drilled bucket down into the middle bucket. Regularly add kitchen scraps and small, fine yard waste like leaves and a small amount of grass clippings. Rotate the buckets when you empty the castings from the middle bucket and the fluid from the bottom bucket.

WORM-KEEPING TIPS

Keep the worms in your vermiculture bin happy, and they'll reward you with plenty of rich, natural fertilizer for the garden—or even your indoor plants.

- Place your vermiculture bin out of direct sunlight as much as possible.

- Add a cup or two of the casting compost from a previous batch when starting a new bucket.

- Harvest worm castings only when you're sure most of the worms have migrated to the bucket above. If you're not sure, remove a pile of castings and leave it outside in the sun. Worms will migrate to the bottom. Remove the top material as it dries out and then return any leftover worms to the top bucket.

1.

Japanese Beetle Trap

Japanese beetles are amazingly damaging pests. The biggest problem is that there are few plants that aren't on their menu. They attack hundreds of different species, laying waste to prize rose beds, destroying emerald-green lawns, devouring strawberries, and much, much more. Although these unwelcome garden visitors are easy to physically catch because they are slow and clumsy fliers, the time and effort it takes means that traps are the preferred method to deal with a Japanese beetle invasion.

Traps like the one in this project are a little bit controversial. Some studies and experts have suggested the trap's bait may draw more beetles into the area than the trap catches. That's why you should put your trap far from target plants, ideally in a corner with only shrubs or groundcover around it. Check the trap regularly—and plants that have been affected by the beetles—and you should notice a diminishing number of beetles on plants as the trap fills up.

The beauty of using such a large container for a Japanese beetle trap is that, unlike smaller types, you can "set it and forget it." Although some homemade traps use homemade baits, usually including a combination of crushed fruit, honey, water, and yeast, this trap uses the tried-and-true pheromone bait. Pheromone baits are widely available at home centers, hardware stores, and nurseries, and are relatively inexpensive. The bait mimics the pest's sex hormone, which is a powerful lure to the Japanese beetle.

Beetles are different from other pests; even as the trap fills up, you'll continue to collect beetles because trapped individuals don't warn off others. Just the same, you need to make sure that trapped beetles don't make their way back out of the bucket, so it's important to add two or three inches of soapy water in the bottom of the bucket. When it's time to dispose of the captured beetles, it's best to do it down a toilet or sewer. If you decide to bury the beetles you've captured and any are alive, they will start their reproductive cycle all over again.

WHAT YOU'LL NEED

Time: **30 minutes** | Difficulty: **Easy** | Expense: **$**

TOOLS:

2" hole saw

Cordless drill and bits

Sharpie (optional)

MATERIALS:

5-gal. bucket with lid

Long-spout funnel (such as an auto-transmission funnel)

Japanese beetle lure (puck or bag style)

Heavy-gauge wire or coat hanger

HOW YOU MAKE IT

1. Cut a 2" hole in the center of the bucket lid with the hole saw.

2. Drill a hole opposite the hanging hole at the top edge of the funnel. Cut the funnel spout about one-third up from the end (to allow for mature beetles to pass easily through the spout).

3. Position the bucket away from affected plants. Add about 3" of soapy water into the bottom of the bucket. Secure the lid on top of the bucket and slip the funnel, spout first, into the hole.

4. Run wire through 1 hole at the top of the funnel and twist the end to secure the wire. Slide the lure onto the wire and secure the wire through the opposite hole, so that the lure is hanging above the mouth of the funnel. Check the trap daily to ensure that it is working where it's located and to determine if it needs to be emptied.

YARD AND GARDEN INNOVATIONS

Upside-Down Tomato Planter

Crazy as it might at first sound, hanging your tomato plants upside down offers a lot of benefits over planting them in the garden. Unlike in the garden, hanging locations usually have abundant direct sunlight because other garden plants aren't competing for the light and shading your tomato plants. Getting the plants up off the ground also gets them away from many diseases and pests that can decimate your delicious crop. A hanging location ensures plenty of air circulation and puts the plants right in the line of sight, where it's easy to detect any potential problems before they get out of hand.

Letting gravity sort things out also means that you don't need to provide support for the plant as it fruits—no cages or stakes. Another big plus is that you water from the top and then water and nutrients are drawn naturally down to the roots. Lastly, say goodbye to weeding with an upside-down planter.

Beyond the practical, upside-down tomato plants are a visually interesting addition that—coupled with a painted and perhaps stenciled bucket—add to the look of the house or yard, wherever they're being hung.

Be aware that topsy-turvy gardening does present a few modest challenges. You have to be absolutely certain that the hook and structure from which you hang the planter can support not only the weight of the bucket when newly planted, but also the weight of a mature plant heavy with fruit and a bucketful of wet soil. Watering can also be difficult for shorter gardeners, and you should be careful to site the planter where water leaking out of the hole in the bottom isn't going to cause a problem.

That said, try this style of gardening with your tomato crop, and chances are that you'll see the opportunity for other edibles. Although you can use this planter for a variety of tomatoes, from heirloom to cherry (it is generally not appropriate for larger varieties such as beefsteak), it's also a wonderful way to grow squash, peppers, cucumbers, herbs, and even mini varieties of eggplant. In fact, you may find yourself creating an entire garden hanging from your porch rafters!

WHAT YOU'LL NEED

Time: **20 minutes** | Difficulty: **Easy** | Expense: **$**

TOOLS:

Sharpie

Utility knife

2" hole saw

Cordless drill and 1" spade bit

MATERIALS:

5-gal. bucket with tight-fitting lid (preferably one with an integral gasket)

Landscaping fabric

5 gal. potting soil

Tomato plant

½" eyebolt

Gallon milk jug or plastic liter bottle

HOW YOU MAKE IT

1. Use the bottom of the bucket as a template to draw a circle with the Sharpie on the landscaping fabric. Cut out the circle with the utility knife. Use the hole saw to make a 2"-diameter hole in the center of the bucket's bottom.

2. Line the bottom of the bucket with the circle of landscaping fabric. Fill the bucket with potting soil, and secure the lid in place, making sure that it's fastened tight.

3. Turn the bucket upside down and use the utility knife to slice an *X* in the landscape fabric covering the bottom hole. Push the root ball of the tomato plant down into the potting soil, and press firmly around the stem of the plant.

4. Screw the eyebolt into the overhang of a shed or the house, or horizontally into a fence post or other solid, secure support. Hang the bucket right-side up by its handle.

5. Cut the milk jug or liter bottle in half crosswise. Use the spade bit to make a hole in the center of the bucket lid. Push the neck of the jug down into the soil. Use the jug to add water to the plant on a regular schedule. As an alternative, run a drip irrigation line into the hole. You can also leave the bucket lid off if you prefer, and the plant may grow stems out of the top (although you risk losing dirt to wind or having animals dig in it).

1.

3.

5.

Vertical Wall Planter

Vertical planting has become widely popular for space- and sun-challenged homeowners (and renters) who want to grow a garden but lack the area in their yard to do it. By mounting planters on a vertical surface, the home gardener can make the most of available light, control the growing medium, and beautify a wall, fence, or other outdoor surface.

There are two primary requirements for a wall planter location—it has to experience at least four to six hours of direct light (unless you've chosen to grow shade-tolerant plants) and it needs to provide stable support. A planter full of wet soil and plants can represent a significant amount of weight, especially as plants mature.

This is a very space-economical way to grow plants, and you can grow almost any garden vegetable in one of these planters. Carrots, bell peppers, squash, tomatoes, and more will all thrive in a vertical five-gallon bucket planter. Not only does the planter hold plenty of nutritious soil, but the design includes handy built-in drainage to ensure your plants never get waterlogged. You can even set this planter up with a drip irrigation system so that you can set it and forget it for maximum ease.

As wonderful as this planter can be for edibles, it's even more so for ornamental plants. Flowers, from miniature roses to begonias, will do quite well, often bulking up to fill the planter with wonderful foliage and blooms. The planter can, amazingly enough, be used for indoor plantings as well. Create a stunning indoor "green wall" with bushing and trailing plants crowded into several of the planters placed side by side on an interior wall. The advantage to using trailing vines and plants with thick, dense foliage is that eventually they will hide the bucket and appear to be growing out of the vertical surface itself—a very magical look. One caveat, though: you may want to forego the drainage holes when growing indoors to prevent any water problems with your indoor flooring.

TOOLS:

Sharpie

Cordless drill and bits

Jigsaw or utility knife

MATERIALS:

5-gal. bucket without lid

80-grit sandpaper

Heavy-gauge wire

Heavy-duty landscape fabric

¼×3" flathead wood screw

HOW YOU MAKE IT

1. Use the Sharpie to mark a cut line that runs from the top edge behind the handles on both sides, down to the midpoint of the bucket on each side, and horizontally across the surface of the bucket.

2. Drill an access hole and then cut along the cut line with a jigsaw. Work slowly to avoid any significant deviations from the line, but don't worry about small errors, because they won't affect the planter in the long run. Sand the cut edges smooth.

3. Drill a random pattern of 10 to 15 holes ⁵⁄₆₄" in diameter in the bottom of the bucket. Drill 3 holes ¼" in diameter, spaced evenly along the top back edge of the planter about 1" down from the top.

4. Use the bucket as a template to cut a circle of heavy-duty landscaping fabric. If hanging from a cyclone fence, use heavy-gauge wire threaded in and out of the holes in the planter's back edge and interwoven with the fence. Loop each end of the wire around the length threaded through the holes and twist it to secure the planter. For a solid surface, use a flat washer with a ¼×3" flathead wood screw, and drive the screw into a support such as a fence post or a wall stud.

5. Line the bottom of the planter with the landscaping fabric, and fill with potting soil. Add plants and water thoroughly.

Strawberry Pot

Strawberries can be a delicate fruit to grow. Susceptible to damage from snails and slugs, they can also ripen on one side if lying on the ground, and conditions have to be just right for the plants to bear the maximum amount of fruit and grow as healthy as possible.

A great way to control growing conditions and maximize the yield is to plant strawberry plants in a strawberry pot. Legend has it that the idea for this space-efficient vertical garden originated as a way to reuse broken terra cotta wine jars in ancient Greece and Italy. Whatever the origins, the idea behind a strawberry pot remains valid: grow the plants up off the ground and you make it harder for pests and diseases to ruin your crop, and easier to make sure your plants are getting everything they need to grow strong.

The pot not only allows you to control the soil and moisture your plants enjoy, but it can also regularly be turned to ensure that all the strawberry plants get the same sun exposure. One advantage to using a plastic five-gallon bucket instead of the more traditional terra cotta is that the plastic will not wick moisture out of the soil. It's also less prone to breakage than terra cotta—a big advantage if you're trying to grow strawberries in a backyard frequented by children. You can always paint it a dusty red-orange to match the look of terra cotta.

Although you may be tempted to remove the handle, it would be wise to leave it attached to the bucket. Strawberries are perennial plants; they will bear fruit for successive seasons if protected during colder months. The handle will allow you to move the bucket into a garage or storage shed after the plants are done bearing fruit, to keep them safe over the winter. It will also make it easier to turn the bucket and get even sun exposure all the way around.

Even if you're not a big fan of growing strawberries, you can use this bucket for small-space gardening. It can accommodate an herb garden or even an ornamental flower or succulent garden. The method of crafting and planting the pot stays the same.

TOOLS:

Sharpie

3" hole saw

Cordless drill and bits

Table saw or jigsaw (or hacksaw)

Vise

Scissors

MATERIALS:

5-gal. bucket without lid

2½" PVC pipe

Landscaping fabric

Crushed gravel, broken terra cotta, or other coarse,

irregular fill

Potting soil

Strawberry plants

Garbage-can lid

80-grit sandpaper

1.

HOW YOU MAKE IT

1. Mark 3"-diameter holes around the outside of the bucket. The pattern of the holes should be staggered, with 2 stacked holes near the top and bottom next to a hole along the middle of the bucket. Drill out the holes with the hole saw. Turn the bucket upside down and drill a random pattern of ⁵⁄₃₂" holes across the bottom.

2. Cut 3" pieces of PVC pipe with the saw. The number of sections should match the number of holes you drilled in the side of the bucket. Cut the sections at a severe angle (to create the bottom of 2 sections), followed by a straight cut (to create the top of the next section). These are cut in the same way as if you were making stakes.

3. Cut a 16" length of the PVC pipe. Secure it in a vise and drill ¼" holes in a random pattern all around the surface of the pipe.

4. Cut a circle of landscaping fabric to fit in the bottom of the bucket. Line the bottom of the bucket with the fabric and add a 2" layer of your fill material. Stand the drilled pipe up in the center of the bucket and pour in potting soil up to the first holes.

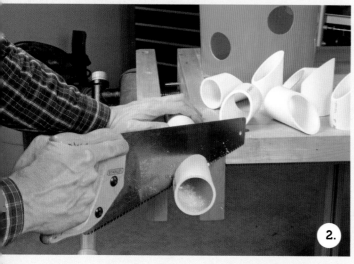

2.

5. Shove a pipe section into one of the lower holes at a downward angle. The long side of the "point" should be on the bottom. Push the pipe section in until only about 1 to 1½" projects out from the bucket and the pointed end is secured in the potting soil. Repeat with the other lower holes.

6. Add more potting soil to fill the bucket to the top of the bottom holes. Plant the strawberry plants in these holes, firming around them with additional potting soil as necessary. Be careful to keep the crown of each plant above the level of the soil.

7. Continue this process of adding soil and planting the plants until you've filled the bucket. Plant more plants in the top as desired. Water thoroughly, in the center watering pipe and in the individual pockets. Place the bucket on the upside-down garbage can lid, on loose rock or gravel, or on bolsters so that the drain holes are not blocked.

PROJECT OPTIONS

Want a bigger haul of delicious, sweet red berries? No problem—this project is scalable. To create a bigger planter with greater yield in the same modest footprint, stack a second bucket—modified with holes and drainpipe—right over the first. Start by cutting the bottom off the top bucket. Drill the holes and "stake" pipes as you did for the bottom planter. Push the cut bottom edge down into the top of the bottom bucket and then plant the top bucket just as you did with the bottom. Then starting enjoying all that luscious fruit.

3.

5.

6.

YARD AND GARDEN INNOVATIONS

Garden Fountain

A basic fountain is one of the most transformative and evocative features you can add to your yard or garden. Fountains large or small provide an appealing visual wherever they stand in the yard, and the sound of burbling water is hypnotic and relaxing (a good argument for placing this fountain close to a bedroom window). These are wonderful for drawing songbirds into the backyard and offer you the chance to grow exotic water-based plants. You can even add a fish or two to your landscape.

The ultimate location of this fountain will be determined to some degree by where the nearest outdoor power outlet is located. However, you can use a long extension cord to increase options for placement. It will also need to be near a water source, such as an outdoor spigot. For aesthetics sake, take pains to conceal the power cord and the hose that feeds the pump, to create the illusion of a natural, standalone fountain (just be sure it doesn't run anywhere it might get damaged).

The project here uses a large garden vase or pot as the "face" of the fountain, with the bucket holding the guts of the system. You can buy a decorative pot, or you may even have one sitting around waiting to be used. You can also paint or stencil a basic terra cotta pot to create a very unique fountain. The fountain could even be run up into a birdbath with a hollow base and a hole in the center. No matter what vessel you use for the fountain, it needs to be smaller at the base than the mouth of the five-gallon bucket on which it will stand. As long as the water will run over the edge, down the pot, and into the bucket, the fountain will work perfectly.

It's a good idea to landscape around the fountain to make it seem like a natural part of your yard and garden. You can also modify the fountain to feed a pond, if you happen to have that alluring garden feature.

WHAT YOU'LL NEED

Time: **1 hour** | Difficulty: **Moderate** | Expense: **$$$**

TOOLS:

Shovel

Torpedo level

Straightedge

Sharpie

Hacksaw

Measuring tape

Jigsaw with metal cutting blade or angle grinder

½" ceramic bit (optional)

MATERIALS:

5-gal. bucket without lid

1" PVC pipe (38" section)

Submersible pond pump

Flexible pond pump tubing

Expanded metal sheet

Plastic sheeting

Cordless drill and bits

Decorative pot or similar container

Silicone sealant

HOW YOU MAKE IT

1. Select a location for the fountain and dig a hole roughly the size and shape of the 5-gal. bucket. Check that the bucket fits in the hole and sits level, the top slightly below the surrounding soil.

2. Remove the bucket and use the straight-edge and Sharpie to mark 3 pairs of cut lines on either side of the bucket, spaced evenly across the mouth of the bucket. Use the hacksaw to cut V slots for the pipes.

3. Measure and cut the PVC pipe into 12½" sections to sit in the slots. Check that the top of each pipe sits level with the top of the bucket's rim.

4. Attach one end of the pump tubing to the pump. Place the pump in the bottom of the bucket with the power cord running out of the top.

5. Measure, mark, and cut a circle of expanded metal sheet 2" larger in diameter than the mouth of the bucket, using a jigsaw with a metal cutting blade or an angle grinder. Put the bucket in the hole, and lay the PVC pipe sections in their slots, with the tubing running between two of the pipes, out of the top. Make sure the power cord is trailing out of the top and away from the bucket (toward the outlet you'll use for the fountain). Lay a donut of plastic sheeting around the mouth of the bucket, and set the expanded metal sheet on top, with the pump tube threaded through the screening.

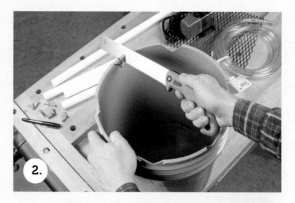

2.

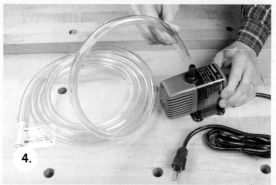

4.

5.

6. Drill a hole in the bottom of the pot you've chosen for the fountain, if there isn't already a drainage hole (or enlarge the hole for the pump tube if it is too small). For terra cotta or ceramic pots, use a ceramic bit and tape over the location before drilling.

7. Run the pump tube up through the hole in the pot, and set the pot in place on the metal sheet, centered over the bucket. Lay a bead of silicone sealant around the base of the tube, and let it dry.

8. Cut a piece of the PVC pipe slightly longer than the width of the pot. Slip the pipe in through the mouth of the pot and wedge it skewed near the top. (If the pot you are using is a perfect cylinder or narrower at the top than the bottom, cut a longer piece of PVC pipe and wedge it from one bottom edge up against the opposite side of the pot.)

9. Wire the pump tube to the PVC pipe so that the mouth of the tube is just below the level of the pot's mouth.

10. Fill the bucket ¾ full of water, through the expanded metal sheet. Fill the pot with water. Plug in the pump and test that the splash of the water out of the tube is at the level you prefer, and that the water running out of the pot is feeding down into the bucket correctly. You can also fill the pot with river rocks for a nice look.

11. Cover the expanded metal sheet and plastic sheeting collar around the bucket with river rock or other decorative stone. Add dirt and plants around the fountain as desired, and make sure the power cord is hidden from view.

9.

THE SIMPLER VERSION

If you don't have a decorative pot or vase for this fountain, you can make a more basic version with a pile of stones and the other materials listed. Paint the inside of your bucket black, and then position it where you want the fountain, with the pump and tube set up (you can support the tube by wiring it to a piece of PVC pipe mounted in a matching PVC flange). Pile rocks around the bucket, filling in with dirt so that the construction looks natural. Conceal the power cord through gaps in the rocks or underneath plant foliage. Fill the bucket to the top with water, plug in the pump, and you'll have a natural-looking fountain.

QUICK 3

Three Quick and Efficient Garden Bucket Options

You don't necessarily have to knock yourself out to turn a five-gallon bucket into a truly useful gardening accessory. Some simple modifications can transform this most basic of containers into simple additions that can benefit any yard or garden.

1.
Rolling Composter

The quickest way to turn garden waste and kitchen scraps into compost is to regularly turn the compost pile. That can take a lot of time and effort. The solution is to roll the pile. High-end drum composters run $100 or more, but you can make your own using a recycled bucket with a tight-fitting lid. Drill eighth-inch holes in a random pattern all the way around the bucket, top to bottom. Fill the bucket with kitchen scraps or other compostables, turn it on its side, and roll it around daily. You'll have usable compost within two weeks. Use additional buckets as needed.

2.
Walking Watering Can

This is perfect for gently watering large areas of a garden or lawn as you walk. You'll lose less water to evaporation than you would with a sprinkler and get exercise in the process. Drill a random pattern of one-eighth-inch holes all over the tight-fitting lid of a clean bucket with a handle. Cut a section of clothesline a few inches longer than the distance from the handle mount to the bottom of the bucket. Loop and knot one end around one side of the handle where it goes into the bucket and the other end on the other side. Add a slip-on handle or wrap the line in masking tape. Fill the bucket half full of water, secure the lid and turn the bucket upside down. Walk along, watering your garden or lawn.

3.
Compost and Soil Sifter

Give your plants the best home possible by ensuring the soil they're in or the compost you feed them is as fine as possible, to maximize moisture retention and nutrient release. Turn a five-gallon bucket upside down and drill the entire bottom with a pattern of 5/16" holes, leaving slim margins of plastic between each hole. This creates a screen that will effectively sift out any large particulates in soil or compost. Add about half a bucket of material, swing it back and forth by the handle, and finer particles will sift out through the holes.

Post-Mounted Birdhouse

If you find the idea of filling your yard with bird song attractive, you can create a home for birds to make them long-term tenants rather than short-term transients. The right birdhouse in the right place will draw in feathered friends who are nice to look at, lovely to listen to, and insect-eating machines.

Before you start building this birdhouse, you'll need to decide on what type of bird you're hoping to entice. Larger birds such as blue jays can be bullies and will take over a nest of smaller birds, if given half the chance. You'll need to figure out the proper entry hole size for the species you want to house. If the holes are too large, larger species and predators will take advantage. And, of course, if the holes are too small, the birds simply can't use the birdhouse.

Different bird species are also particular about other elements of their housing, including how close or far away the birdhouse is kept from people and how high up it will be situated. Placement is critical in any case. Birds, like other creatures, want to be safe from predators in their homes. This means keeping the birdhouse at a safe remove from bushes or other structures that cats could use as cover as they stalk their feathered prey. Make your decisions based on a "bird's-eye" view of things and you won't go wrong. Do your research and the birds you're looking for will find the home irresistible.

This project is the perfect opportunity to develop your plastic painting and stenciling skills. You can paint or decorate your birdhouse bucket with designs that help it blend in, or something more distinctive to make it a focal feature. However, it's a good idea not to attach any decoration to the bucket or post; the slick surfaces are important to help keep predators off the birdhouse and out of the birds' lives.

TOOLS:

Measuring tape

Sharpie

Cordless drill and bits, including 1" spade bit (or size that corresponds to the hole size for the birds you want to attract)

Circular saw or hacksaw

Paintbrush

MATERIALS:

5-gal. bucket with lid

PVC cement

½" PVC flange

½×6" threaded PVC nipple

3×60" PVC pipe

3" PVC slip flange

PVC primer

¾" plywood scrap at least 15" square

Sandpaper

1" flathead wood screws

Oil-based primer and paint

Quick-setting cement

(2) small metal L brackets or plastic shelf supports

3.

HOW YOU MAKE IT

1. Measure and mark two holes spaced evenly along the length of the bucket, top to bottom. Use the spade bit to drill out the holes (choose a hole size based on the size of the type of bird for which the house is meant).

2. Cut the inside of the lid out to the size that will allow it to be slid down into the bucket, as close to exactly in the middle of the bucket as possible. Test that the disc fits securely, and measure from where the disc sits to the top edge of the bucket.

3. Use PVC cement to glue a threaded ½" PVC flange to the cut disc, centered on the disc. (You can also screw the flange to the inner floor with small nuts, bolts, and fender washers for added stability, and larger bird species.) Screw a ½×6" threaded PVC nipple into the flange. Secure the disc in the bucket with the nipple sticking out of the top. Mark and cut the nipple so that the end is level with the rim of the bucket. (If the nipple is too short, use a longer nipple and cut it as necessary.)

4. Coat the end of the 3×60" PVC pipe and the inside of the 3" PVC slip flange with PVC primer. Let the primer cure according to the manufacturer's directions, and then coat the surfaces with PVC cement and attach the flange to the end of the pipe. (You can use a longer or shorter pipe, depending on how high you want the birdhouse to stand.)

5. Use the mouth of the bucket, with the cut lid rim in place, as a template to trace a circle on a piece of ¾" exterior grade plywood (or you can use hardwood if you have a large, square scrap on hand). Drill an access hole and cut around the marked circle, leaving about ½ to 1" extra all the

way around. (See page 22 for instructions on cutting a circle out of plywood). Sand the plywood circle, and prime and paint with oil-based white paint, or the color you prefer.

6. Screw the plywood circle onto the post's slip flange, using the 1" screws. The flange should be centered on the circle.

7. With the divider shelf in place, place the cut lid rim on top of the bucket. Hold the post upright and plumb (you'll need a helper or a sturdy workmate for this), and sit the bucket upside down on top of the plywood circle. Make sure it is centered.

8. Drill pilot holes down through the bucket's lid rim and into the plywood. Screw the rim to the plywood with screws.

9. Dig a hole 14 to 16" deep. Pour in quick-setting cement and add water. Mix the cement with a scrap piece of wood, poking it to release any air holes. Stand the birdhouse post in the hole and check for plumb. Once you're sure that the post is plumb, screw braces of scrap wood to either side of the post and secure it in place with these temporary braces until the cement dries.

10. When the cement cures completely, decorate the pole and/or birdhouse as you desire. Fasten the L brackets or plastic shelf supports underneath the entry holes for the birds to land on. You can also drill small holes and stick short sections of dowel under the entry holes if you prefer a slightly cleaner look. Consider adding a light plastic pot with annual flowers to the top of the birdhouse so that potential predators cannot perch on top of the house.

5.

6.

8.

Aeroponics System

A lack of usable yard space or a paved-over backyard aren't excuses not to grow a bountiful, edible garden. Aeroponics is the practice of growing plants without soil, usually in a very space-efficient way. It's actually just a version of hydroponics that involves crowding plants together and giving the roots exactly the nutrients and water they need to thrive. The process is all about efficiency. By controlling the growing medium in this way, gardeners can optimize the harvest from their plants with a minimum of time, effort, space, or resources.

The wonderful thing about this particular aeroponics setup is that it's easy and adaptable. It won't even take an hour of your time to set up, and it can be used indoors or out for a wide range of plants. Although most people use aeroponics for edibles or consumable crops, you can also use it propagate ornamental plants such as roses or even shrubs. It's a wonderful way to get any plants off to a great start, and to begin growing your garden before the weather outdoors is conducive to growth.

Studies have also shown that the right aeroponics system can increase the harvest of certain plants. Researchers have found that tomato plants grown this way sometimes yield twice the amount of tomatoes as plants grown in a more conventional manner. The same is true for other species and varieties.

You'll find the project that follows can easily be adapted to the plants you want to grow. Some will need larger pots, but that isn't a problem. Just make sure you buy an appropriate nutrient solution for the plants you want to grow, and replenish it according to the label instructions. You can find nutrient solutions through larger and specialty garden retailers, some nurseries, and online. Also keep in mind that many municipal water systems add chlorine to their water. If you're using tap water, it's recommended you let it stand for at least an hour before adding it to the aeroponics system. That should be plenty of time to allow the chlorine to evaporate.

TOOLS:

Cordless drill and bits, including 1" spade bit

Sharpie

Hole saw (sized to match diameter of the pots you're using)

Circular saw (or hacksaw)

Timer

MATERIALS:

Food-grade 5-gal. bucket with lid

1¼" rubber grommet

300-gal./hr. hydroponics pump with ½" threaded fitting

½×10" threaded riser

½" 360° mister nozzle

Net pots (or gravel)

Nutrient solution

HOW YOU MAKE IT

1. Use the spade bit to drill a 1" hole near the top of the bucket, just below the lowest ridge on the side.

2. Mark the holes for the net pots on top of the lid with a Sharpie (use a pot as a template). Cut the holes out with the hole saw.

3. Screw the riser into the pump. Install the mister on the riser. Set the pump in the bucket.

4. Run the power cord out through the hole; pull it taut, so that there isn't extra cord coiled inside. Slide the rubber grommet over the cord end and secure it in the hole. Add water to just over 2" above the pump. Run the pump for a moment to ensure the system is operating correctly. Add the nutrient solution.

5. Secure your plants in the pots with the lids (or use a growing medium). Set the pots in place, making sure they're pushed all the way down into the lid.

6. Plug the pump into the timer, and the timer into an outlet. Set the timer for 30-minute intervals. Check for the first 2 hours that the timer is working correctly and the plant roots are getting sprayed sufficiently.

2.

3.

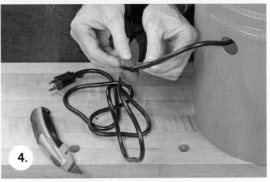

4.

Index

Resources

Allway Tools
(718) 792-3636
allwaytools.com
Manufacturer of replacement lids
for paint-filled five-gallon buckets
that include a pour spout

Berkley
(800) 237-5539
www.berkley-fishing.com
Manufacturer of a fishing rod
holder for five-gallon buckets

Big Bear Products
(269) 657-3550
www.bigbearproducts.com
Supplier of a five-gallon bucket
lid-replacement seat, called the
Silent Spin Seat

Bucket Boss
(888) 797-7855
bucketboss.com
Supplier of accessories for
five-gallon buckets to be used on
construction and contracting sites

Doulton Ceramic Filter Systems
(800) 664-3336
www.doulton.com
Supplier of candle-style ceramic
filters for use in five-gallon-bucket
filtration systems

Gamma 2
www.gamma2.net
Producer of the Gamma Seal Lid,
which replaces the stock bucket lid
with a tightly sealed screw-on lid

J-B Weld
(903) 885-7696
www.jbweld.com
Producer of epoxy products

Krylon
(800) 457-9566
www.krylon.com
Producer of spray paints for
plastics

Oneida Air Systems
(800) 732-4065
www.oneida-air.com
Maker of the Dust Deputy, a
cyclonic separator that is attached
to a modified five-gallon bucket to
catch dust from power tools

OrigAudio
(949) 407-6360
www.origaudio.com
Maker of a stick-on amplifier that
can turn a five-gallon bucket into
a music-player speaker

Original Bucket Dolly
(631) 256-5888
www.originalbucketdolly.com
Producer of a caster dolly meant
for use with five-gallon buckets

Vestil Manufacturing, Inc.
(800) 348-0868
vestilmfg.com
Manufacturer of stiff plastic pail
liners for five-gallon buckets

Woodcraft Supply
(800) 225-1153
www.woodcraft.com
Manufacturer of a five-gallon
bucket lid replacement to convert
the bucket into a shop cyclonic
dust collector

About the Author

Chris Peterson is a freelance writer and editor based
in Ashland, Oregon. He has written extensively on
home improvement and general reference topics,
including books in the Black & Decker Complete
Guide series; *Building with Secondhand Stuff: How
to Re-Claim, Re-Vamp, Re-Purpose & Re-Use
Salvaged & Leftover Building Materials*; *Practical
Projects for Self-Sufficiency: DIY Projects to Get
Your Self-Reliant Lifestyle Started*; and *Manskills:
How to Avoid Embarrassing Yourself and Impress
Everyone Else*. When he's not writing or editing,
Chris spends his time snowboarding on the
mountains of the Pacific Northwest and rooting
for the New York Yankees.